Lecture Notes in Economics and Mathematical Systems

472

Springer
Berlin
Heidelberg
New York
Barcelona
Hong Kong
London
Milan
Paris
Singapore
Tokyo

Jean-Robert Tyran

Money Illusion and Strategic Complementarity as Causes of Monetary Non-Neutrality

Springer

Author

Dr. Jean-Robert Tyran
University of St. Gallen
Department of Economics
Bodanstr. 1
CH-9000 St. Gallen, Switzerland

Library of Congress Cataloging-in-Publication Data

Tyran, Jean-Robert, 1967-
Money illusion and strategic complementarity as causes of monetary non-neutrality / Jean-Robert Tyran.
p. cm. -- (Lecture notes in economics and mathematical systems, ISSN 0075-8442 ; 472)
Includes bibliographical references and index.
ISBN-13: 978-3-540-65871-9 e-ISBN-13: 978-3-642-46883-4
DOI: 10.1007/978-3-642-46883-4
1. Money--Mathematical models. 2. Monetary policy--Mathematical models. I. Title. II. Series.
HG220.5.T96 1999
332.4--dc21 99-23349
CIP

ISSN 0075-8442
ISBN-13: 978-3-540-65871-9

Typesetting: Camera ready by author
SPIN: 10699885 42/3143-543210 - Printed on acid-free paper

Acknowledgements

In principle, money illusion could provide an explanation for the inertia of nominal wages and prices and, thus, for the non-neutrality of money. Yet, the notion of money illusion seems to be thoroughly discredited in economics. TOBIN (1972), for example, described the negative attitude of most economic theorists towards money illusion as follows: "An economic theorist can, of course, commit no greater crime than to assume money illusion." As a consequence, money illusion has been anathema to mainstream economics since the 1970s. Rather than referring to money illusion, economic theorists aimed at building macroeconomic models from the assumption that all individuals are rational and hold rational expectations. This rational expectations revolution certainly was a theoretical success since it allowed the building of consistent and elegant models of macroeconomic interaction. Yet, the empirical success of this enterprise remains questionable because tests of these rationality assumptions and their implications were fraught with difficulties.

I am grateful to Professor ERNST FEHR (Zurich) who encouraged me to question fundamental assumptions and current attitudes in the economics profession. He also taught me not to commit "crimes", i.e. not be content with *assuming* that people are prone to money illusion. Instead, one should *observe* what people really do under controlled conditions. ERNST FEHR taught me experimental economics which is the prime method to test fundamental assumptions and their implications in economics.

Several people at the Institute for Empirical Research in Economics (Zurich) supported me. SIMON GÄCHTER's critical comments motivated me to try to improve. I thank him, as well as MARTIN BROWN for comments on the manuscript and for proof reading. URS FISCHBACHER provided the software for the experiment (an early version of Z-Tree) and made helpful suggestions to the experimental design. BEATRICE ZANELLA, TOBIAS SCHNEIDER and MARTIN BROWN helped me to recruit the experimental subjects and to conduct the experiments. LORENZ GÖTTE supported me with the layout of the book.

I received much social support from my parents, friends and the people at the Institute. CAMILLA CONSIGLIO patiently accompanied me through all the highs and lows of the last few years. I thank all these people for their encouragement.

Contents

Introduction

"The work for which I have received the Nobel Prize was part of an effort to understand how changes in the conduct of monetary policy can influence inflation, employment, and production. So much thought has been devoted to this question and so much evidence is available that one might reasonably assume that it had been solved long ago. But this is not the case: It had not been solved in the 1970s when I began my work on it, and even now the question has not been given anything like a fully satisfactory answer."

ROBERT E. LUCAS (1996: 661)

Why can changes in monetary policy cause fluctuations in employment and production? This has been one of the most important and most intensely debated questions in economics ever since DAVID HUME's (1752) contributions. The issue of monetary non-neutrality is important because the misconduct of monetary policy may create enormous damage to the real economy and thereby affect the economic lives of millions of people. For example, abrupt and massive reductions in the quantity of money are claimed to have been responsible for severe depressions like the Great Depression in the 1930s or the 1982 recession in the United States. Thus, investigating the causes of monetary non-neutrality is important because it may help avoid misconduct in monetary policy. Despite the immense amount of theoretical and empirical research that has been produced in the last two centuries, the issue remains intensely debated for two reasons. First, it is difficult to identify and accurately measure the variables of interest. For example, the quantity of money is a concept which is not easily defined, and the price level is difficult to measure. In addition, presumed causalities are hard to establish by analyzing field data since many (unobservable) variables may change simultaneously. The second reason why the question of monetary neutrality remains controversial is of methodological nature. Economists have established a firm tradition to approach the issue of monetary (non-) neutrality by *assuming* that all economic agents are fully rational and form rational expectations. This assumption has kept its strong position up to the day. Hence, money illusion has been dismissed as an explanation of monetary non-neutrality on *a priori* grounds.

The present study uses the methods of experimental economics to investigate whether money illusion and strategic complementarity are causes of monetary non-neutrality. To identify these causes, the adjustment of nominal prices and its effect on real incomes is studied after an anticipated monetary shock in an experimental environment without exogenous frictions. Experimental methods are particularly suitable for this task because they allow the accurate observation of human behavior under controlled conditions and to isolate causes of monetary non-neutrality.

The main finding of this study is that money illusion has been prematurely dismissed as an explanation of monetary non-neutrality. This result is based on theoretical considerations and empirical evidence. On the theoretical level, it will be argued that to generally rule out aggregate effects of money illusion, extremely restrictive rationality assumptions have to be made. On the empirical level it will be shown that non-neutrality after an anticipated shock is the rule rather than the exception. Most importantly, this effect will be shown to be caused by the interaction of money illusion and strategic complementarity.

The book contains four Parts:

Part I discusses the different approaches to the problem of monetary non-neutrality that have been chosen in the literature. It explains why it is difficult to decide on empirical grounds using field data whether money is neutral or not. Most theories aiming at an explanation of nominal rigidity and the non-neutrality of money start with the assumption that economic agents are fully rational and hold rational expectations. Monetary non-neutrality is then explained by introducing exogenous frictions. However, evidence from economic psychology indicates that the assumption of full rationality is restrictive. In particular, money illusion seems to be pervasive at the individual level and may therefore be potentially relevant in many fields of economics. On the theoretical level it is explained how the interaction of money illusion and the strategic properties of the economic environment may explain monetary non-neutrality. Various hypotheses, arising from New Classical macroeconomics, economic psychology and game theory are formulated.

Part II presents an experimental design that allows the testing of the proposition of New Classical macroeconomics that an anticipated monetary shock is neutral in the absence of exogenous frictions. It explains what the specific advantages of the experimental method in investigating monetary non-neutrality are. These include the exact measurement of variables like prices and real incomes, the observability of expectations, and the control over incentives and informational conditions. The latter allows the implementation of a truly exogenous and anticipated monetary shock in a friction-free environment, and to discern whether realizations of endogenous variables and expectations are in or out of equilibrium. Above all, experiments allow the isolation of truly causal factors which is not possible with field data. Five treatments have been designed to isolate causes of monetary non-neutrality.

Part III presents the results of this experimental study. The first major finding is that the New Classical proposition on the neutrality of anticipated monetary shocks can clearly be rejected: When the environment is represented in nominal terms and strategic complementarity prevails, money is massively non-neutral in the short run. The postulate of long-run neutrality, on the other hand, receives confirmation in the data. Analysis of expectations data shows that aggregate real income losses occur because agents do not form rational expectations.

Subsequent chapters isolate causal factors of short-run monetary non-neutrality by controlled *ceteris paribus* variations of the economic environment. The results show that if more nominal elements are included in the representation of the economic environment, agents are less capable to "pierce the veil of money" and more pronounced non-neutrality is observed. Thus, money illusion is a causal factor in explaining monetary non-neutrality when the environment is characterized by strategic complementarity. When the environment is characterized by strategic substitutes, money is approximately neutral in the short run. Thus, the New Classical proposition of neutrality of anticipated money is a reasonably good predictor of aggregate behavior in the case of strategic substitutes. Furthermore, it is shown how money illusion *interacts* with strategic properties of the economic environment: Money illusion matters more with strategic complements than with strategic substitutes, and the strategic properties matter more when money illusion can be assumed to be prevalent.

Part IV discusses the results of this experimental study. First, some issues related to the experimental design are addressed. This experimental study is part of an effort to make macroeconomics an indirectly experimental science. Obviously, the decision environment for the individual in the laboratory differs in many respects from the decision environment in an actual macro-economy. It is certainly much simpler than that of a central bank governor or even of the man in the street. However, this is one of the strengths of the experimental approach: by controlling the environment, we are able to test the behavioral relevance of assumptions underlying standard macroeconomic theory. Second, the findings of this study are discussed in a broader context, and conclusions for macroeconomic theory and policy are drawn. The policy conclusions, however, are necessarily speculative in nature. Unfortunately, the results of this study provide no clear-cut and direct guidance for the appropriate conduct of monetary policy. Rather, the study suggests caution. If (simple) macroeconomic theories are unable to explain behavior under the (simple) conditions implemented in the laboratory, one should be very sceptical that they will succeed in doing so in the much more complicated naturally occurring economy. It is hoped that the findings of this study motivate theorists to build macroeconomics on firmer behavioral foundations.

Part I

Approaches to the Problem of Monetary Non-Neutrality

[There is a discord between] "two intuitive feelings about money which are held simultaneously by most economists most of the time. The first is that, at least in the long run, 'money does not matter' to the real economy, i.e. to the determination of relative prices, output and employment. [... The idea is that] one must 'try from the start to pierce the monetary veil in which most business transactions are shrouded' [...]. The second intuition is that 'money does matter', at least in the short run, i.e. that inappropriate management of the money supply (a concept not easily defined) can result in serious damage to the economy ... through real effects on prices and output."

EATWELL *et al.* (1989: xi)

"'Neutrality of money' is a shorthand expression for the basic quantity-theory proposition that it is only the level of prices in an economy, and not the level of its real output, that is affected by the quantity of money which circulates in it." (PATINKIN, 1989: 273). The opposite, "monetary non-neutrality", means that changes in nominal money affect real economic variables, such as output and employment. Investigating the problem of monetary non-neutrality thus implies investigating the question of how the monetary and the real side of the economy interact. Empirical and theoretical knowledge about these interdependencies is fundamental to the conduct of economic (monetary) policy, since (as EATWELL *et al.* state above) if money is non-neutral "inappropriate management of the money supply ... can result in serious damage to the economy".

As the above quotation from EATWELL *et al.* (1989) suggests, there are two basic intuitions about the effect of money on the real economy. The intuition of monetary neutrality is inspired by basic (microeconomic) theory, the intuition of monetary non-neutrality is inspired by (macroeconomic) empirical findings. Yet, the evidence on whether changes in the quantity of money do or do not affect real variables remains controversial (see chapter 1). Why can diametrically opposed opinions on the non-neutrality of money persist despite the immense amount of empirical literature? There are basically two reasons: first, there may be considerable mismeasurement in aggregate variables which seriously affects the conclusions drawn from empirical work with macroeconomic data; second, it seems to be impossible to discriminate empirically among competing theories on the basis of available macroeconomic data and to establish truly causal relations among aggregate variables (see section 1.2.).

If one believes, as many economists and practitioners do, that "money matters" at least in the short run, how can this be explained theoretically? According to elementary economic reasoning, as incorporated in the quantity theory of money,[1] changes in nominal money should be neutral, if all economic agents are rational and all markets work perfectly.[2] The usual approach to explain why monetary non-neutrality arises is the concept of **nominal rigidity** (or nominal inertia). This term captures the phenomenon that nominal variables (like prices and wages) do not immediately adapt to a change of some monetary aggregate, but adjust in a "sluggish" or "sticky" manner. Consequently, in order to account for the non-neutrality of money, one has to explain nominal rigidity. Traditionally, theorists started from the assumption of fully rational agents (having rational expectations) and focused either on informational or on market imperfections ("frictions") of some kind to explain nominal rigidity. Theories of nominal rigidity based on the assumption of fully rational agents are discussed in section 2.1.

However, the above statement that changes in money should be neutral if all agents are rational and all markets work perfectly offers a second natural theoretical starting point for the investigation of monetary non-neutrality. Instead of relaxing the assumption that all markets work perfectly all of the time, one could just as well *relax the assumption of full rationality* of all actors. However, this approach has been deemed unacceptable by most economists. Some argue that the assumption of individual rationality should not be given up for *a priori* (methodological) reasons and is to remain the cornerstone of any economic and, thus, of monetary theory also. Others argue that even if some agents were irrational sometimes, this would not matter on an aggregate level. Despite this scepticism, some theorists have recently begun to explore the theoretical implications of *small* deviations from full rationality. Such

1. See FRIEDMAN (1989) for a survey and discussion of the quantity theory of money.
2. DAVID HUME (1752) "has deduced the quantity theory of money by purely theoretical reasoning from 'that principle of reason' that people act rationally and that this fact is reflected in market-determined quantities and prices." LUCAS (1996: 664).

deviations may stem from agents who do not have fully rational expectations (e.g. because it is costly to form such expectations), or who do not perfectly maximize (e.g. because losses from errors are only of second order). Instead of looking at small deviations from full rationality, one could alternatively consider large deviations from full rationality by very *few* agents. As discussed in section 2.2. these theories suggest that small deviations from rationality may under some circumstances lead to large business cycle fluctuations. The theoretical argument of the present study builds on HALTIWANGER and WALDMAN (1985, 1989). These authors argue that few not fully rational agents disproportionately affect the aggregate when the environment is characterized by **strategic complementarity**, but have a disproportionately small impact when the environment is characterized by strategic substitutes.

Chapter 3 discusses evidence from the economic psychology literature. In particular, the study by SHAFIR, TVERSKY and DIAMOND (1997) suggests that many people have problems to discern nominal from real variables correctly and that even if they are informed about the difference they continue to actually prefer some nominal changes over others for a given real change. Thus, the evidence from economic psychology suggests that people seem to be prone to **money illusion**. Or, to refer to the initial quotation: a lot of people seem to have problems "to pierce the monetary veil in which most business transactions are shrouded." It is argued that people tend to think in money terms because money is a natural and salient unit. IRVING FISHER suggested that money illusion is caused by a confusion which results when money loses its function as a reliable measure of economic transactions. Modern research explains money illusion as a framing effect. First, people tend to be nominally "anchored", i.e. to judge changes from some naturally (e.g. through experience or custom) given nominal starting point. Second, people may have problems to distinguish correctly between nominal and real income ("high-number illusion"). Based on the evidence from economic psychology, one may speculate that money illusion could be potentially relevant in many fields of economics (section 3.2.). Though suggestive, most of the evidence from economic psychology research stems from questionnaire studies. Economists tend to be sceptical about this kind of evidence resulting from hypothetical questions. Rather, economists are interested in the actual and interactive behavior of people motivated by economic incentives.

Chapter 4 provides the theoretical background for the experimental study. It is argued that the rationality requirements for ruling out monetary non-neutrality, even in an environment without exogenous frictions, are extremely restrictive in a pricing game. In particular, one has to assume common knowledge of the absence of money illusion to rule out that money illusion matters in the aggregate. It is explained how money illusion may affect expectations and, depending on the strategic properties, behavior. The considerations of this chapter allow the provision of more precise notions of why money illusion may be a cause of monetary non-neutrality. In particular, we distinguish between "first-order money illusion" which directly affects behavior, and "higher-order money illusion" which affects behavior through its effect on expections.

Chapter 5 summarizes Part I and states the hypotheses for the experimental study. The Null hypothesis is that an anticipated monetary shock should be neutral in a friction-free environment. There are three alternative hypotheses. The first states that the aggregate effects of money may arise because people are led to take suboptimal actions because of money illusion. This hypothesis states that "money illusion matters", independently of strategic properties. The second alternaitive hypothesis states that aggregate effect of monetary shocks arise because people (for some unexplained reason) are heterogeneous with respect to rationality. In the presence of this type of heterogeneity, strategic properties matter (independently of money illusion). The third alternative hypothesis combines the other two and states that money illusion and strategic properties interact in a specific way.

Chapter 1: Empirical evidence on the non-neutrality of money from macroeconomic data

"In the short run, which may be as long as three to ten years, monetary changes primarily affect output. Over decades, on the other hand ... the rate of monetary growth primarily affects prices.... One major finding has to do with severe depressions. There is strong evidence that a monetary crisis, involving a substantial decline in the quantity of money, is a necessary and sufficient condition for a major depression."

MILTON FRIEDMAN (1989: 32)

"The observation that money changes induce output changes in the same direction receives confirmation in some data sets but is hard to see in others. Large-scale reductions in money growth can be associated with large-scale depressions or, if carried out in the form of a credible reform, with no depression at all."

ROBERT E. LUCAS (1996: 668)

" ... the evidence concerning whether monetary shocks have important real effects is controversial"

DAVID H. ROMER (1996: 241)

This chapter provides a very brief survey of the empirical literature investigating monetary neutrality using macroeconomic time-series data. The main purpose of this chapter is to show that the issue of monetary non-neutrality is controversial and to provide some explanations for this state of affairs.

Section 1.1. presents some selected evidence concerning the following questions: Is money procyclical? Do changes in money cause short-run changes in real aggregates? Is money neutral in the long run? Are anticipated changes in money neutral?

Section 1.2. discusses some problems which arise when investigating these questions.

1.1 Selected evidence on the (non-)neutrality of money

The empirical debate on the neutrality of money is still ongoing. Most empirical studies find a procyclical movement of money aggregates, though some contest even this view (KYDLAND and PRESCOTT, 1990). Even when accepting the result that monetary aggregates are procyclical, one is not in the position to conclude from this that money is non-neutral (FRIEDMAN and KUTTNER, 1992, 1993). Procyclicity may just as well stem from a third factor influencing both money and real aggregates or from effects from the movement of real aggregates on monetary aggregates. The extent of this "reverse causation" still is empirically debated (COLEMAN, 1996). Conventional wisdom interprets the empirical relation between monetary aggregates and measures of real aggregate economic activity primarily as reflecting the effect of monetary policy on real activity. A host of historical episodes is apparently in accord with this interpretation.

Some authors claim to be able to show empirically that recessions are caused by restrictive monetary policy. For example, ROMER and ROMER (1989) argue that the 1982 recession in the USA was caused by restrictive monetary policy under Fed-chairman VOLCKER. Another example is provided by BERNANKE and CAREY (1996) who purport to show that the Great Depression in the 1930s has been caused by a contraction in money and nominally rigid wages.

SARGENT (1976) interpreted the prediction that *anticipated* money would have no real effects, as the hypothesis that money would *not cause*, in the sense of GRANGER (1969) and SIMS (1972), changes in unemployment rates. He found that this hypothesis was confirmed for U.S. time series data. On the other hand, BALL (1991) concludes that output growth is on average below normal following announced shifts to tighter money.

BARRO (1977) decomposed money growth into two components, one unanticipated, obtained as the residual from a forecasting equation, i.e. regressions of M1 on its own lagged values, and one anticipated and equal to the forecast value. BARRO concludes that the hypothesis that only the unanticipated component affected unemployment and output could not be rejected.[3] MISHKIN (1983) re-examined the data, and concluded that both unanticipated and anticipated components have long-lasting effects on output. However, the decomposition of money has also been questioned

(KING, 1981).

SARGENT (1982) claims to be able to show that large, sudden reductions in the rates of money growth (the monetary and fiscal reforms that ended four of the post World War I European hyperinflations) were *not* associated with output reductions that were large by historical standards, or possibly by any depressions at all. This finding stands in stark contrast to the quotation by FRIEDMAN (1989: 32) above. SARGENT goes on to argue that these reductions in money growth rates were well anticipated which may be the reason for the neutrality of money in these cases.

The central predictions of the quantity theory of money are that, in the *long run*, money growth should be *neutral* in its effects on the growth rate of production and should affect the inflation rate on a one-for-one basis (see FISHER and SEATER, 1993; BOSCHEN and OTROK, 1994). MCCANDLESS and WEBER (1995) check simple correlations of inflation rate and money growth over a 30 year (1960 to 1990) horizon. Using 110 country data sets, they report very high correlations.[4] LUCAS (1996: 666) comments this finding: "It is clear from these data (and from many other studies that have reached similar conclusions) that the applicability of the quantity theory of money is not limited to currency reforms and magical thought experiments." On the other hand, BLANCHARD (1990: 828) argues that "long-run neutrality of money ... is very much a matter of faith, based on theoretical considerations rather than on empirical evidence."[5]

3. ENDERS and FALK (1984) replicate BARRO (1977, 1978) with a microeconomic test of monetary neutrality (they investigate the market for U.S. pork and claim to avoid aggregation problems thereby). For studies who refine the approach by BARRO, and provide evidence in favor on the New Classical hypothesis that anticipated money is neutral, see also KRETZMER (1989) or KORMENDI and MEGUIRE (1984).
4. These range from 0.92 to 0.99, depending on the monetary aggregate chosen and the number of countries included.
5. In fact, "empirical tests of long-run neutrality are often difficult to interpret, since assumptions usually have to be made about the underlying structure of the economy." (OLEKALNS, 1996: 393). Alternatively, one has to assume that the money supply is truly exogenous.

1.2. Problems of empirical work in monetary macroeconomics

The preceding section has shown that the evidence concerning whether monetary shocks have important real effects is controversial. Why is there so much disagreement on this question? Section A) argues that there may be substantial mismeasurement in macroeconomic aggregates which may importantly affect the conclusions drawn in empirical studies on monetary neutrality. Section B) discusses fundamental problems of isolating causality in monetary economics.

A) Mismeasurement in macroeconomic aggregates

In empirical research it is often difficult to decide which variables should be considered in the first place. One reason for this is that macroeconomic variables are generally aggregate variables which cannot be observed directly, but have to be constructed using data from different sources. This construction of aggregate variables relies on some (implicit or explicit) theory of measurement and aggregation. Prominent examples of problems arising in measurement and aggregation are *price indices*. The determination of the price level is necessary for calculating real quantities (like real GDP or real money) from observed (or, rather, constructed) nominal quantities. However, measuring the price level is not unproblematic. SHAPIRO and WILCOX (1996) discuss the probable amount of mismeasurement in the consumer price index (CPI) in the U.S. They find that the upward bias of the CPI is centered on *1 percentage point per year*. However, the extent of this bias is not known exactly (see MOULTON, 1996).

BELONGIA (1996) provides a striking example of how sensitive macroeconomic conclusions may be with respect to measurement problems. Specifically, he discusses how inferences of the effects of money on economic activity may depend importantly on the choice of *monetary indices*. BELONGIA replicates five recent studies on real effects of money. In four of the five cases, the qualitative inference in the original study is reversed when a simple-sum monetary aggregate is replaced by a Divisia index of the same asset collection. The economic justification for not using simple-sum indices (like M2) is that aggregating any set of commodities with equal weights implies that each good is a perfect substitute for every other good in the group. Yet, this condition seems to be strongly rejected by empirical evidence. The problem with simple-sum aggregates is that they cannot internalize pure substitution effects, and are thus prone to spurious shifts that would suggest a change in the utility derived from money holdings, when, in fact, no such change has occurred. The studies replicated concern the quantification of monetary shocks, the symmetry of money's effect on output, the relationship between money and the business cycle, and money-income causality.[6]

In all, mismeasurement may be an important explanation of the contradictory results found in empirical studies on monetary non-neutrality.

B) Problems of isolating causalities

Why is the debate between Real Business Cycle theory (which poses that purely monetary disturbances have no real effects) and Keynesian theories (where monetary changes have important effects on output) so difficult to decide on empirical grounds? (see ROMER, 1996: 232-6). Couldn't we just regress real GNP on M1 with different lags?[7] Unfortunately, regressions of that kind do not provide any evidence in favor of monetary theories and against real theories. The basic reason is that money may not be exogenous for three different reasons.

First, a major difficulty in testing empirically for the monetary effects on real economic activity is that the money supply not only influences economic activity but also is influenced by it in turn. Causation may run from output to money rather than the other way around. This reverse causation may result from shifts in money demand stemming from changes in firms' and households' production plans (KING and PLOSSER, 1984). As a result, we may see changes in the money stock in advance of output movements even if the changes in money are not causing the output movements. In this case lagged money will help predict output even if it does not affect it (for an early discussion see TOBIN, 1970; for a summary, see BLANCHARD, 1990).[8]

---

6. The replicated studies are: ROTEMBERG (1993), COVER (1992), KYDLAND and PRESCOTT (1990), FRIEDMAN and KUTTER (1992), and STOCK and WATSON (1989) versus FRIEDMAN and KUTTER (1993).
7. As in the famous St. Louis Equation, see ANDERSEN and JORDAN (1968).
8. To give an example of nonsensical results one can arrive at, when blindly applying statistical causality tests, consider the following question (after PINDYK and RUBINFELD, 1991: 218f.): Which came first: The chicken or the egg? THURMAN and FISHER (1988) have finally shed some light on this issue by using causality tests. They use annual data on two variables: total U.S. production of eggs from 1930 to 1983 and total U.S. production of chickens. To conclude that one of the two "came first", it is necessary to find unidirectional causality, i.e. to reject the noncausality of one to the other and at the same time fail to reject the noncausality of the other to the one. Using different lags, the authors obtained a clear rejection of the hypothesis that eggs do not cause chickens, but were unable to reject the hypothesis that chickens do not cause eggs. Thus they were able to conclude that the egg came first!

Second, monetary policy may interact with movements in other aggregates. Suppose monetary authorities adjust the money stock to try to offset other factors that influence aggregate output. If this monetary policy intervention proves to be successful (i.e. it has real effects), we will observe movements in money but not in output. Similarly, suppose fiscal and monetary policies are coordinated (e.g. both are expansionary). In this case we may observe a strong correlation between movements in money and output, even if money has no effect at all on real economic activity. Hence, we cannot conclude from the positive correlation between money and output that money causes output, neither can we conclude that money does not affect output if we fail to observe such a correlation.

Third, financial innovations and deregulation of financial markets in the last few decades may have led to large shifts in money demand. If monetary authorities do not adjust money supply to these demand disturbances, we may observe a negative correlation between money and output. As a result of such money demand shifts, the estimated relationship between money and output is very sensitive to such matters as the sample period and the measure of money.

ROMER (1996) concludes from these problems that even more sophisticated statistical analyses of the association of money and real variables will not provide strong evidence concerning the relative merits of monetary and real theories of fluctuations. One way to arrive at meaningful empirical evidence is to look for "natural experiments" (e.g. ROMER and ROMER, 1989). The most famous example of this approach is the work of FRIEDMAN and SCHWARTZ (1963) which undertook a careful historical analysis of the sources of movements in the money stock from 1867 to 1960. Based on this historical analysis they claim to be able to discern which movements in the stock of money stem from the monetary sector, and which from real developments. This claim rests on the assumption that the monetary intervention under study was truly exogenous, and has not in turn been caused by some real economic factor. Since the exogeneity of monetary interventions can never be shown to hold beyond any doubt, it remains questionable whether these authors have in fact shown the monetary contraction to be the *cause* of monetary non-neutrality.

Chapter 2: Theories of nominal rigidity and monetary non-neutrality

"...departures [from full employment due to an anticipated change in the money supply] simply cannot occur in any model that assumes ***fully*** *rational optimizing behavior, including rationally formed expectations, unless nominal prices exhibit inertia. The logic is simple. Rational agents should only care about real magnitudes. If so, any optimizing model with a unique equilibrium will be 'money neutral'. A cut in the supply of money should just cause a proportional reduction in prices and wages with no change in employment, output, or real wages. This conclusion in no way depends on the assumption that markets are perfectly competitive, or even that markets clear."*

GEORGE A. AKERLOF and JANET L. YELLEN (1987: 138)

This chapter presents a short overview of theories of monetary non-neutrality which rest on the assumption that monetary non-neutrality arises because nominal prices do not adjust immediately after a monetary shock.[9]

Section 2.1. provides a short review of theories of nominal rigidity that start from the assumption of fully rational agents. In this case, monetary non-neutrality is explained by referring to information problems, a combination of imperfect competition and small cost of adjusting prices, or prices that have to be set in advance.

Section 2.2. discusses theories that explain nominal rigidity by relaxing the assumption that all agents are rational all the time.

9. We are thus not concerned with the specific "transmission channels" (e.g. through interest rates, exchange rates, asset prices or credits), which all assume the presence of nominal rigidity at some stage of the chain of argumentation. See MISHKIN (1995) for a concise summary.

2.1. Explaining monetary non-neutrality and nominal rigidity by assuming that all agents are fully rational

"We have a wide variety of theories that reconcile long-run monetary neutrality with a short-run trade-off. They all ... carry the implication that anticipated money changes will not stimulate production and that at least some unanticipated changes can do so."

ROBERT E. LUCAS (1996: 667)

The issue of monetary non-neutrality has been of central interest to economists for at least 200 years. In this very long period many outstanding researchers have made important contributions to the field. An attempt to review this literature is beyond the scope of this book and would lead us too far afield. Rather, some important contributions which relate to the issue under study have been selected and are briefly discussed below. For a more extended discussion see PATINKIN (1989), BLANCHARD (1990), MANKIW (1990), MANKIW and ROMER (1991), ANDERSEN (1994) or ROMER (1996: Ch. 6). The following discussion heavily draws on these surveys.

Until the early 1970s, the New Classical synthesis, a consensus view of macroeconomics which emerged in the 1950s and 1960s, did not treat price and wage decisions explicitly. Prices were assumed to be markups over costs, and wages were explained by the augmented PHILLIPS curve. Most research on wage and price behavior was until then characterized by a strong empirical bent and a rather eclectic use of microeconomic justifications (BLANCHARD, 1990). This broad consensus faltered in the early 1970s because of two weaknesses (MANKIW, 1990): One theoretical and one empirical. The empirical weakness was that the stable relation between unemployment and inflation (the PHILLIPS-Curve) seemed to disappear in the mid 70's.[10] The theoretical weakness was that the theories were not firmly rooted in sound microeconomic theory. The following revolution of providing "microeconomic foundations of macroeconomics" in the 1970s was largely a theoretical one and was mainly concerned with modeling expectations properly. The general position was accepted that expectation formation should be viewed as *rational expectations*.

From this revolution, two important lines of theorizing emerged: the New Classi-

10. However, the debate on the PHILLIPS-Curve received new impulses in the 1990's. For example, the Winter-issue of the *Journal of Political Economy* [1997: 11(1)] is devoted to this question.

cal and the New Keynesian. Both approaches contributed to the understanding of the microfoundations of macroeconomics, and both used the assumption of fully rational actors with rational expectations.[11] The New Classical macroeconomics combines the assumption of rational expectations with permanently clearing, perfectly competitive markets. In this context money can only be non-neutral, if agents have an informational problem (a "signal extraction problem") due to the fact that they only observe own prices. According to this view, only unanticipated changes in money are non-neutral, whereas *anticipated changes are completely neutral.*[12]

The Real Business Cycle approach is related to the New Classical approach. It proposes that business cycle fluctuations are mainly caused by (real) shocks to preferences and technology, whereas changes in money are thought to be irrelevant (i.e. neutral).

Three ways have been explored to model nominal frictions: First, in a perfectly competitive economy, nominal rigidity arises because producers do not observe the aggregate price level. This approach works under the maintained "as if" assumption of perfect competition in all markets but relaxes the assumption of full information. Second, small costs from changing nominal prices (menu costs) or some other small friction in nominal adjustment may lead to real effects from monetary changes. Third, in the models of staggered adjustment, monetary shocks have real effects because not all prices or wages are adjusted simultaneously.

Relaxing the assumption of perfect information. The central idea of the LUCAS-PHELPS[13] model is that when a producer observes a change in the price of his product, he does not know whether it reflects a change in the good's relative price or a change in the aggregate price level. A change in the relative price alters the optimal amount to

---

11. MANKIW (1990: 1658): "... the axiom of rational expectations is as firmly established in economic methodology as the axioms that firms maximize profits and households maximize utility."
12. This statement is obviously only an approximation for naturally occurring economies. BLANCHARD (1990: 780): "Any anticipated change in nominal money must lead to anticipated changes in the price level, and thus introduce a wedge between the opportunity cost of holding money and the cost of capital; in all cases this will affect utility and, in most cases, is likely to affect capital accumulation as well ...Even unanticipated changes, if they are the result of open market operations, are likely to be non-neutral: open market transactions will usually involve some but not all holders of money and have distribution effects... But except for the case of steady inflation which may be substantial (especially when the non-neutrality of the tax system is taken into account), these effects are mere intellectual curiosities; ... For that reason, most of the research has taken as given that prices do not adjust fully and instantaneously to nominal money and focused on the reasons for and implications of imperfect price adjustment."
13. LUCAS (1972) and PHELPS (1970).

produce. A change in the aggregate price level, on the other hand, leaves optimal production unchanged. When the price of the producer's good increases, there is some chance that this reflects a rise in the good's relative price (signal extraction problem). The rational response for the producer is to attribute part of the change to an increase in the price level and part to an increase in the relative price, and therefore to increase output somewhat. This implies that the aggregate supply curve is upward sloping: when the aggregate price level rises, all producers erroneously view this as an increase in the relative price of their goods, and thus raise their output. In accordance with this approach, and under the assumption that markets always clear, LUCAS claimed that only unanticipated changes in the quantity of money will have real effects. An *anticipated* change in the quantity of money will be expected by the individual to affect all prices proportionately (LUCAS, 1972, 1973, 1975). Although there is a statistical output-inflation relationship in this model, there is no exploitable trade-off between inflation and output. If policymakers attempt to take advantage of statistical relationships, effects operating through expectations may cause a breakdown of these relationships. This is the famous LUCAS critique (LUCAS, 1976).

Relaxing the assumption of competitive markets. It is not sufficient to move from the assumption of perfect competition to imperfect (e.g. monopolistic) competition to explain nominal rigidity. Even assuming that markets do not clear (e.g. because of unions, implicit contracts etc.) is not sufficient for nominal rigidity to arise (see the above quotation by AKERLOF and YELLEN, 1987: 138). Thus, just assuming that an economy is characterized by *real rigidities* (as in efficiency wage models), is not sufficient to generate nominal rigidity and the non-neutrality of money. Yet, as explained below, the combination of real rigidities with (exogenous) nominal frictions (like menu costs) or with the assumption of boundedly rational agents is sufficient to explain nominal rigidity and the non-neutrality of money.

Relaxing the assumption of costless price adjustment. MANKIW (1985) discusses a model with monopolistic goods markets and small *menu costs*, i.e. cost from changing nominal prices. He shows that if all firms are in equilibrium, individual losses to firms who do not adjust nominal prices after a nominal shock may be small (second order), but the welfare effects of this non-adjustment may be large (first order). The idea behind this argument is basically the envelope theorem, i.e. the intuition that "hills are flat at the top". A weakness of the menu cost argument is that it assumes that all prices are initially in equilibrium. If prices are not all equal or optimal to start with, it is no longer obvious that even small changes in nominal money will leave all prices unaffected.

The argument of BLANCHARD and KIYOTAKI (1987) is based on a combination of pecuniary externalities and menu cost. In a model of monopolistic competition, a decrease in an individual producer's nominal price has two effects: First it increases the demand for that producers good, second it increases real money balances (slightly), thus increasing the demand for all other producers. This latter effect is called *aggregate demand externality.* Because aggregate output is initially below its socially optimal level in monopolistic competition, this effect will in the presence of small menu cost lead to an increase in welfare after a positive monetary shock. Imperfect competition implies that, in response to an increase in nominal money, the incentive to adjust relative prices may be weak.[14] Small costs of changing prices will prevent adjustment of relative prices, thus of nominal prices, leading to an increase in aggregate demand.

Relaxing the assumption of permanently flexible prices in price setting. The so-called "staggering" models of FISCHER (1977) and TAYLOR (1979) showed that rational expectations could be introduced in the wage-price-mechanism, but nevertheless generate long-lasting monetary non-neutrality. The simplest *time dependent* rules are such that the time between price decisions is fixed. An alternative staggering structure is considered in CAPLIN and SPULBER (1987). These authors derive the aggregate behavior of prices and output in response to changes in nominal money when individual price setters follow *state-dependent* rules. These rules imply that the nominal price is adjusted whenever the difference between the actual price and the target price exceeds some fixed threshold value. Their result shows that menu costs do not necessarily imply non-neutrality of money. Thus, the static menu cost argument does not extend straightforwardly to the dynamic case. With time dependent rules, menu costs generate real effects of nominal money. With state-dependent rules, money may still be neutral.

14. Hence, an important property of this model is strategic complementarity. This property will be discussed in more detail in chapter 4 of this Part.

2.2. Explaining monetary non-neutrality by assuming that some agents are not fully rational

"... rather than assuming rational expectations and then introducing imperfections, why not start with the more realistic assumption that agents are heterogeneous in terms of their information processing abilities?"

JOHN HALTIWANGER and MICHAEL WALDMAN (1985: 336)

The main bulk of the literature aiming to explain the causes of nominal rigidities rests on the assumption that all agents are fully rational. The New Keynesian approach *assumes* some kind of exogenous nominal friction or non-indexed contract. This approach is not very satisfying and seems to be quite close to the old Keynesian assumption of nominally rigid wages. It would be much more satisfying to show that nominal rigidities can arise *endogenously* and one way to do this is to relax the assumption of fully rational agents.[15]

If all or most economic agents suffered from money illusion, money would obviously not be neutral. This assumption has in general been discarded because it seemed to most economists all too simple and "ad hoc". The economics profession preferred (and still seems to prefer) to show that monetary non-neutrality may arise under the maintained *as if*-assumption that all agents are fully rational. However, some theorists have abandoned the assumption of full rationality. These theorists have chosen the following procedure: They attempt to show theoretically that *small*, exactly defined deviations of *few* actors from full rationality may under plausible conditions imply large business cycle fluctuations.

EVANS and RAMEY (1992) argue that the assumption of rational expectations is a misspecified model, since it does not account for the (deliberation or opportunity) cost of forming such expectations. The authors show that if fully optimizing agents have to bear small costs of expectations formation money cannot be neutral.

AKERLOF and YELLEN (1985a, 1985b) construct a model where firms can set prices in monopolistically competitive goods markets and set wages in an efficiency-wage labor market. If a small fraction of firms acts in a "near-rational" manner, nomi-

15. In the literature menu costs are interpreted as costs that are *external* to the decision maker (e.g. the cost of actually printing new menues in a restaurant). A bounded rationality interpretation could be that such menu cost are *internal* to the decision maker. If it takes time and effort to take a decision, menu costs could also be interpreted as opportunity cost of time.

nal wages and prices adjust sluggishly after a monetary shock. An agent is said to behave in a near-rational manner, if he or she ignores second order losses from small deviations of optimality. Simulations of their model show that large non-neutrality may arise under plausible parameter values (see also NAISH, 1993).

HALTIWANGER and WALDMAN (1985, 1989) show in a model with heterogeneous agents (i.e. a large number of agents is fully rational and but a few agents are not fully rational) that the strategic property of the economic environment is decisive to determine whether nominal rigidities arise and money is non-neutral or not. The authors show theoretically that, if the environment is characterized by strategic complementarity, a small number of boundedly rational agents have a disproportionate impact on aggregate behavior. If, on the other hand, the environment is characterized by strategic substitutes, the few boundedly rational agents have a disproportionately small impact on aggregate behavior.[16] The concept of strategic complementarity has many highly important interpretations in naturally occurring economies (OH and WALDMAN, 1994). Examples are monopolistic competition, economies of scale in production, search activities in labor markets, and (positive) spillovers between macroeconomic aggregates or markets can essentially be captured by the concept of strategic complementarity. BOMFIM and DIEBOLD (1997) have shown in a macroeconomic simulation study that heterogeneity with respect to rationality may cause money to be non-neutral under reasonable parameter values.

The above studies show theoretically that a small relaxation of the assumption of full rationality can potentially have a large impact on the aggregate behavior of the economy. The empirical relevance of these studies critically depends upon the validity of their assumptions concerning the boundedness of individual rationality. An explicit experimental testing of these assumptions and their empirical impact on the aggregate outcome of real economic agents is a central task of the present study.

16. The concepts of strategic complements and strategic substitutes are discussed in chapter 4.

Chapter 3: Money illusion

"This paper begins with the premise that theory which fits the real world will be based on assumptions that individuals are not fully rational. It would be simply unscientific to proceed otherwise. For indeed, individuals may actually suffer from money illusion ... in economic matters."

GEORGE A. AKERLOF and JANET L. YELLEN (1987: 138)

In principle, money illusion could provide an explanation for the inertia of nominal prices and wages and, thus, for the non-neutrality of money. The stickiness of *nominal* prices and wages seems to be an important phenomenon (see section 3.2.B) and has puzzled economists for decades because it is quite difficult to explain in an equilibrium model with maximizing individuals. Yet, the notion of money illusion seems to be thoroughly discredited in modern economics. TOBIN (1972), for example, described the negative attitude of most economic theorists towards money illusion as follows: "An economic theorist can, of course, commit no greater crime than to assume money illusion." As a consequence of this negative attitude, money illusion is anathema in mainstream economics. For example, the index of the handbook of monetary economics (FRIEDMAN and HAHN, 1990) does not even mention the term "money illusion". Instead of referring to a concept so alien to mainstream economics, researchers have sought for explanations which are based on rational agents holding rational expectations. Factors like informational frictions, costs of price adjustment and staggering of contracts have been invoked to explain nominal inertia (see chapter 2.2.). The present study does not contest the potential relevance of these explanations. However, it is argued that money illusion has been prematurely dismissed as a potential candidate for the explanation of sluggish nominal price adjustment. Our argument is based on theoretical considerations and on empirical evidence. At the theoretical level it will be argued that in order to rule out the relevance of money illusion it is not sufficient that individuals are illusion-free but that the absence of money illusion is common knowledge (see chapter 4). Yet, considering the evidence from economic psychology, it seems highly unlikely that this common knowledge requirement is met in practice. At the empirical level it will be shown that, after a fully anticipated negative nominal shock, nominal inertia and monetary non-neutrality is the rule rather than the exception (see Part III).

Why should economists be interested in money illusion at all? There are four reasons why economists should be interested in money illusion. First, there is a considerable amount of evidence from the economic psychology literature that money illusion is an important phenomenon at the individual level. This discipline has convincingly shown that individual decisions can be systematically biased by the *representation* of a decision problem (see section 3.1). Second, if we accept the notion that money illusion is a pervasive phenomenon in individual decision making, it can be hypothesized that outcomes in many economically relevant instances may be substantially affected by this psychological phenomenon. These instances not only include individual decisions (like a portfolio allocation decision), but also specific markets (like labor markets) and macroeconomic aggregates (see section 3.2). Third, recent contributions to economic theory suggest that even very few agents suffering from such illusions may disproportionately affect market or aggregate outcomes in some environments. This, of course, is of central interest to economists. Not many economists would deny that agents sometimes make (even systematic) mistakes. However, the relevant question (for macroeconomics) is whether these individual mistakes have an impact at the aggregate level.

As will be argued in Parts II and III, there is a fourth reason to take money illusion serious: It has proven to be an important factor in explaining monetary non-neutrality in a controlled experiment.

Definitions of money illusion. The term "money illusion" is commonly used to describe any failure to distinguish monetary from real magnitudes. Clearly, different authors have used the term "money illusion" with slightly differing meanings. The term seems to have been coined by IRVING FISHER, who defined it as a "failure to perceive that the dollar, or any other unit of money expands or shrinks in value" (FISHER, 1928: 4).[17] The intuition behind this statement is the following: People tend to use some *smallest nominal accounting unit* (e.g. one US-dollar) as a yardstick to measure economic transactions. If the quantity of money doubles, the same dollar bill as before will only have half of its real value in a frictionless world. According to FISHER, people have problems to understand that the nominal yardstick has shrinked in real terms. FISHER's definition of money illusion concerns a confusion about the real value of

17. See PATINKIN and STEIGER (1989) for an extended discussion on the origin of the terms "veil of money" and "neutrality of money". See MONTESANO (1981) on the notion of "money illusion".

money units in a changing environment. We will henceforth call this confusion FISHER effect. As noted by HOWITT (1989), however, the definition has evolved over time. Many writers have used the term money illusion as synonymous to a violation of what LEONTIEF (1936) called the *homogeneity postulate*; the postulate that demand and supply functions be homogeneous of degree zero in money prices and in the initial quantity of financial assets, including money. That is, these functions depend only on relative prices, but not on the absolute price level. For the necessary and sufficient conditions that must be satisfied by the utility function in order to generate such illusion-free demand functions, see HOWITT and PATINKIN (1980) and DUSANSKY (1980).[18] PATINKIN (1949) used a slightly different definition that also takes into account the potential effects of people's real wealth on their supply and demand behavior. According to PATINKIN money illusion is absent if individuals' net demand functions are homogenous of degree zero in all money prices *and* real wealth. Although the definition of PATINKIN differs from LEONTIEF's definition by taking into account the "wealth constraint", both definitions are based on the same intuition. This intuition says that if the *real* incentive structure, i.e. the *objective* situation an individual faces, remains unchanged, the *real* decisions of an illusion-free individual do not change either. Two crucial assumptions underlie this intuition: First, the objective function of the individual does not depend on nominal but only on real magnitudes. Second, people perceive that purely nominal changes do not affect their opportunity set. For example, people have to understand that an equiproportionate change in all nominal magnitudes leaves the real constraints unaffected. Whether people are indeed able to pierce the veil of money and to understand that purely nominal changes leave their opportunity set unchanged is, in principle, an empirical question. The next section presents evidence to this question.

18. RUSSELL and THALER (1985: 241) illustrate the link between the concept of rationality and money illusion by the following quotation: "SAMUELSON (1983) considered the following problems to seek for violations of the weak axiom of revealed preference. Suppose that we confront an agent with an income-price vector (Y, p) and observe the choice x. Now confront the agent with an income-price vector (mY, mp) where m is a positive constant. Unless the consumer again chooses x, the weak axiom is being violated. The reason is simple. By multiplying both income and prices by m we do not change the budget set. Thus any choice $y \neq x$ at (mY, mp) violates the axiom." Thus, SAMUELSON seems to propose to test money illusion as a framing effect.

3.1. The psychology behind money illusion

" ... we must abandon meta-arguments about whether it is 'possible' that psychologists have identified economically relevant departures from rationality, self interest, and other familiar assumptions. Of course it is possible, and in fact it is true."

MATTHEW RABIN (1996: 3).

One of the remarkably few papers on money illusion was written by the two psychologists ELDAR SHAFIR and AMOS TVERSKY together with PETER DIAMOND, an eminent (macro-)economist (henceforth quoted as STD, 1997). The next two sections heavily draw on the work of these authors. STD propose a psychological account of money illusion. They explain money illusion in terms of multiple representations (different framings) and nominal anchoring.

As already mentioned in the last section, two preconditions are necessary for money illusion not to affect (individual) behavior. First, people's preferences should not be affected by purely nominal changes. Second, people should understand that their opportunity set is not affected by purely nominal changes. STD provide evidence that frequently one or both of these preconditions are violated. Their results suggest that people's preferences as well as their perceptions of opportunities are affected by nominal values. Moreover, many people do not only seem to suffer from money illusion; they also *expect that other people's preferences and decisions are affected by money illusion*. Problem 1 of STD's questionnaire study neatly illustrates these claims. STD presented the following scenario to two groups of respondents:

> *Consider two individuals, Ann and Barbara, who graduated from the same college a year apart. Upon graduation, both took similar jobs with publishing firms. Ann started with a yearly salary of $ 30,000. During her first year on the job there was no inflation, and in her second year Ann received a 2% ($ 600) raise in salary. Barbara also started with a yearly salary of $ 30,000. During her first year on the job there was a 4 % inflation, and in her second year Barbara received a 5% ($ 1500) raise in salary.*

Respondents of group 1 were then asked the happiness question: *"As Ann and Barbara entered their second year on the job, who do you think was happier?"* 36 percent thought that Ann was happier while 64 percent believed that Barbara was happier. This indicates that most subjects believed that preferences are affected by nominal variables because in real terms Ann does of course better than Barbara. Respondents of group 2 were asked the following question: *"As they entered their second year on the job, each received a job offer from another firm. Who do you think was more likely to leave the present position for another job?*" In line with the response to the happiness question 65 percent believed that Ann, who is doing better in economic terms, is more likely to leave the present job. Thus, a majority believed that other people's decisions are affected by money illusion.

Money illusion as the result of framing effects. Since the absence of money illusion means that an individual's preferences, perceptions and, hence, individual choices of real magnitudes are not affected by purely nominal changes it is natural to view money illusion as a framing or representation effect. From this viewpoint an individual exhibits money illusion if the preferences or the perceptions of constraints and the associated decisions depend on whether the same environment is represented in nominal or real terms. Respondents may have based their answers to the scenario above on *high-number illusion*. High-number illusion implies that the homogeneity postulate is violated. If people's nominal incomes double and all nominal prices double, people may fail to perceive that their real opportunity set remains unchanged. In addition, they may actually prefer the higher nominal income. For example, respondents may have believed that Barbara's nominally high raise of $1500 is worth more in real terms than Ann's nominally low raise of $600. However, when explicitly asked, most respondents seem to understand that Ann in fact does better in real terms.[19] This suggests that respondents believe that other people actuall prefer the higher nominal income even though it is smaller in real terms. STD's analysis is based on a large body of research in cognitive psychology that shows that alternative representations of the same situation may well lead to systematically different responses (TVERSKY and KAHNEMAN 1981, 1986).[20] Representation effects seem to arise because people tend to adopt the particular frame that is presented and evaluate the options within this frame. Because some options loom larger in one representation than in another, alternative framings of the same options can give rise to different choices. With respect to

19. There was a third group of respondents who was asked whether Ann or Barbara are doing better in economic terms. 71 percent answered that Ann is in fact doing better in economic terms.
20. See RABIN (1996) for a survey. Examples of systematically differing responses in questionnaire studies include the choice among risky prospects which are either framed in terms of gains or of losses (KAHNEMAN and TVERSKY, 1979) or choice of treatment for diseases like lung cancer (MCNEIL *et al.*, 1988) which are either framed in terms of mortality or survivor rates. Framing has also been shown to affect behavior in voluntary contribution mechanism experiments (ANDREONI, 1995). For an early demonstration of a framing or representation effect in experimental economics see SELTEN and BERG (1970).

money illusion, STD explain that people tend to have multiple representations contemporaneously but the nominal representation is often *simpler and more salient.* STD show that people are generally aware of the difference between nominal and real values, but because money is a salient and *natural unit,* people often think of transactions in predominantly nominal terms.[21] Thus, in the case of money illusion, people's judgments do not correspond to either the real or the nominal evaluation but, rather, to a mixture of the two.

Money illusion as the result of anchoring effects. STD (1997) also relate money illusion to the psychological effect of "anchoring", i.e. to judge changes from some naturally (e.g. through experience or custom) given starting point. Individuals could either use anchors in nominal or in real terms. Again, the saliency and naturalness of nominal (i.e. money) values may be a reason why people frequently use nominal anchors.[22] Obviously, anchoring is related to FISHER's intuition about the confusion arising from using a smallest nominal accounting unit (see page 23). Elderly people are sometimes heard to complain that everything is "so expensive nowadays".[23] This confusion may arise because a relation between a nominal accounting unit and a real unit (e.g. a loaf of bread) is mentally anchored. STD speculate that this may be relevant for housing markets, where sellers may anchor on the historical price that they paid for the house and may be reluctant to sell the house for less than the nominal anchor (see section 3.2. for further examples).

Reliability of questionnaire studies. The inferences in STD are predominantly based on results from questionnaire studies. These studies have the advantage that they allow to investigate preferences and motives. On the other hand, questionnaire studies have obvious limitations. Do answers given to a hypothetical question extend to actual behavior in a real-world context, when incentives are present? Control in questionnaire studies may be insufficient, since subjects may bring to bear other, unspecified assumptions (e.g. from their own personal experience). Interestingly, many

21. In the working-paper version, SHAFIR *et al.* (1996: 35, footnote 18) cite an amusing example of another interesting domain in which nominal-real confusions may arise: thinking about time (which is certainly full of paradoxa). When the Gregorian calendar was adopted in England in 1752, omitting 11 days so that the day ensuing to September 2nd was September 14th, "much discontent was provoked among uneducated people who imagined that they were being defrauded of the omitted days; and there were riots with the cry 'Give us back our 11 days'".
22. For example, AKERLOF and YELLEN (1987: 140) write: "The most natural explanation of sticky money wages stems from anchoring."
23. Or, as YOGI BERRA puts it: "A nickel ain't worth a dime anymore."

findings in behavioral economics that first have been illustrated in questionnaire studies have proven to be surprisingly robust in later experiments (FRANCIOSI *et al.*, 1995; BENARTZI and THALER, 1995; GRETHER and PLOTT, 1979; GRETHER, 1980). In both studies the deviations from full rationality were at least as strong in a condition with monetary incentives as in a condition with purely hypothetical questions. These findings suggest that evidence from questionnaire studies should not be easily dismissed.

3.2. Potential relevance of money illusion to economics

"Now ordinary experience tells us, beyond doubt, that a situation where labor stipulates (within limits) for a money-wage rather than a real wage, so far from being a mere possibility, is the normal case. ... It is sometimes said that it would be illogical for labor to resist a reduction of money wages but not resist a reduction of real wages. ... But whether logical or illogical, experience shows that this is in fact how labor behaves."

JOHN MAYNARD KEYNES (1936: 9).

Most economists are not *per se* interested in whether agents make mistakes in answering questionnaires or in individual decision making. Many (applied and macro-) economists will only be prepared to take the issue seriously if they can be convinced that money illusion matters at the market or even the aggregate macroeconomic level. It is an important purpose of the experimental study presented in Parts II and III to investigate under which conditions this may be the case.

The presence of money illusion has been invoked to account for the short-run non-neutrality of money and for business cycle fluctuations, as in the case of FISHER (1928). On the other hand, monetary theorists have reacted adversely to explanations based on such illusions, partly because money illusion contradicts the maximizing paradigm of microeconomic theory and partly because invoking money illusion is often viewed as an "ad hoc"-explanation of phenomena that do not fit well into the standard equilibrium mould of economics (HOWITT, 1989). This section presents some rather tentative and speculative conjectures on where money illusion *could potentially* be relevant for economics, again drawing on SHAFIR, TVERSKY and DIAMOND (STD, 1997). This section discusses the potential importance of money illusion A) as a disequilibrating force, B) in labor markets and C) in other areas of the economy.

A) Money illusion as a disequilibrating force

In the past economists frequently used the assumption of money illusion to account for the short-run non-neutrality of money. Irving FISHER's explanation of business cycles is, for example, based on lenders' money illusion during an upswing.[24] However, since the success of the rational expectations revolution an extreme reluctance to invoke money illusion as an explanation of the short-run non-neutrality of money has been established. While New Classical macroeconomists focus on informational frictions to account for short-run non-neutrality, New Keynesians mainly focus on costs of price adjustment or staggering (see section 2.1. for references). In the absence of menu costs, staggering, and informational frictions, the models of New Keynesian and New Classical economists rule out that purely monetary changes have real effects. A common feature of these models is that they *exclusively* focus on the equilibrium states of their economies. In general, they remain silent on how economic agents move from one equilibrium to the other. In models that exclusively focus on equilibrium the assumption of the absence of money illusion is very intuitive because it is difficult to imagine that an illusion could persist in equilibrium. However, as will argued next, there is a strong *a priori* argument that money illusion is likely to affect the *adjustment process* of an economy after a fully anticipated monetary shock. The argument is based on the simple fact that a Nash equilibrium involves the coordination of expectations (for an extended discussion, see chapter 4). This can be illustrated in the context of a monopolistically competitive economy as analyzed in, for example, AKERLOF and YELLEN (1985) or BLANCHARD and KIYOTAKI (1987). To keep the argument simple we focus only on firms' behavior. The reduced form real profit function for firms in these models can be written as $\pi_i = \pi_i(P_i/\bar{P}, M/\bar{P})$, where π_i is firm *i's* real profit, P_i is the nominal price set by firm *i*, $\bar{P}$ is the aggregate price level and *M* denotes the supply of money.[25] In these models $M/\bar{P}$ is proportional to real aggregate demand. For simplicity, we assume identical firms and a unique symmetric equilibrium $P_i^* = P_j^*$, for all *i*, *j*. In this

---

24. FISHER believed that lenders are willing to supply more in the face of a rise in nominal interest rates although real interest rates decline or remain unchanged due to inflation.
25. The profit function already incorporates (*i*) the maximizing behavior of all households, (*ii*) the cost minimizing behavior of all firms for given output and wages levels, (*iii*) the equilibrium real wage, and (*iv*) the equilibrium relation between real aggregate demand and real money balances. In AKERLOF and YELLEN (1985) the real wage is given by the Solow condition because firms are efficiency wage setters. In BLANCHARD and KIYOTAKI (1987) housholds are wage setters so that firms take real wages as given when choosing nominal prices and output.

equilibrium each firm maximizes real profits by setting $P_i^* = \bar{P}^*$. Since the profit function is homogeneous of degree zero in P_i, $\bar{P}$ and M it is obvious that a change in M to λM, $(\lambda \neq 1)$, leads to post shock equilibrium values of λP_i^* and $\lambda \bar{P}^*$.

Suppose now that there are agents who believe that there are other agents who suffer from money illusion and do not fully adjust their nominal prices to λP_i. The first group of agents, therefore, anticipates a change in real aggregate demand $M/\bar{P}$ so that their members, in general, have an incentive to choose a price that differs from λP_i^*. For this conclusion to hold, it is not even necessary that there are indeed firms which believe that others suffer from money illusion. Suppose, for example, that there is one group of firms, which believes that a second group of firms believes, that there is a third group which suffers from money illusion and does, hence, not adjust fully. This means that the first group believes that the second group does not choose the equilibrium price λP_i^* and, hence, the first group also has an incentive to choose a price which differs from the equilibrium price. The basic message of this argument is, thus, that unless the *absence of money illusion is common knowledge*, there will, in general be no coordinated instantaneous adjustment to λP_i^*. As a consequence, the economy will go through a process of disequilibrium.

B) Labor markets

Until the early 1970s, many economists had the intuition that money illusion may be relevant in labor markets (see quotation from KEYNES, 1936: 9 above). This intuition has been inspired by the observation that labor markets exhibit large quantity movements and small price movements. Yet, the observation of nominally sticky wages on an aggregate level is not clear evidence for money illusion.[26] For example, BERNANKE and CAREY (1996) argue that the monetary crisis in the U.S. translated into the Great Depression of the 1930s with unprecedented mass unemployment because of nominally rigid wages. Yet, the authors admit (1996: 4) that their finding "leaves open the deep question of why wages did not adjust more quickly". In fact, many of the observations of nominally rigid wages in the aggregate may be explained by theories of nominal rigidity based on the assumption of fully rational economic agents (see section 2.1). On the other hand, aggregate nominal wages may exhibit

26. Several studies based on time-series macroeconomic data investigate whether nominal values affect labor supply decisions. Some of these studies do find indirect evidence of money illusion, others do not (FAIR, 1971; NIEMI and LLOYD, 1981. For an approach using microeconomic data see GUSTAFSON and HADLEY, 1989).

considerable flexibility because of fluctuations of the work force (e.g. when entrants into the labor market earn less than insiders in the business cycle downswing). Thus, some authors have argued that one has to look at *evidence from individual level data*. Recently, data from individually identifiable workers have been used to argue that downward nominal wage rigidity is pervasive (AKERLOF *et al.*, 1996; KAHN, 1997[27]). These studies are motivated by the "fair wage-effort" hypothesis (AKERLOF, 1982, 1990). This hypothesis proposes that given incompletely specified and enforceable labor contracts, workers may reduce their on the job effort if their remuneration is perceived to be unfair.[28] Suppose there is a link between nominal wages and workers' perception of *how fair* they are being treated by their employer. That is, suppose that fairness judgments are affected by money illusion.[29] If this is the case, a fully rational profit maximizing firm anticipating the link between money illusion and perceived fairness will be reluctant to cut nominal wages. What are the consequences if this hypothesis holds? First, contractive monetary policy would lead to a recession with falling prices but nominally stable (i.e. rising in real terms) wages. A second implication of this theory is that there may be benefits of positive, but low rates of inflation which would be "greasing the wheels of the economy" during times of "normal" macroeconomic activity. According to this view, this is due to the fact that structural change is facilitated. Contracting firms and industries should reduce their real wages. With positive inflation rates this is possible by holding nominal wages constant and letting inflation erode the real wage. With zero inflation, these unlucky firms would have to cut nominal wages, which - according to the fair wage-effort hypothesis - would lead to additional real losses.

There is considerable evidence which renders this chain of argumentation plausible. First, we have some evidence on the link of fairness perceptions with nominal wages. Second, we know from questionnaire studies that employers are reluctant to cut nominal wages because they are afraid that working morale suffers and that workers will be less loyal to the firm. The evidence on these issues is now briefly discussed.

27. KAHN (1997: 1006) concludes from her study: "This pattern of wage stickiness implies clear money illusion in its focus on nominal rather than real values."
28. For experimental evidence on the fair wage-effort relation, see FEHR, KIRCHSTEIGER and RIEDL (1993, 1994).
29. AKERLOF and YELLEN (1987: 140) write: "... in the discussion of *fairness*, people's views of fair money wages apparently are anchored in the current *money* wage."

Do workers judge nominal wage cuts as unfair? KAHNEMAN, KNETSCH and THALER (1986) address the role of money illusion in judgments of fairness. Subjects had to judge a company which decreases wages in a no-inflation environment by 7%, as well as a company raising wages by 5% in an environment with 12% inflation. Although the change in real income is approximately the same in both situations, the percentage of respondents who judged the action of the company "unfair" was 62% in the case with the nominal cut but only 22% in the case of the nominal rise. Apparently, judgments of fairness seem to be based largely on nominal rather than on real changes. In LOEWENSTEIN and SICHERMAN (1991) people preferred (nominally) rising wage profiles over "flat" wage profiles even when they were told that the two have the same (real) current present value. STD (1997) conducted questionnaire studies with bywalkers at public places and with undergraduate students. The first series concerned people's attitudes on salary raises (for an example, see page 25). The results indicate that people's perception of well-being is dependent on relative wages (i.e. a social comparison motive) and on the nominal versus real representation. Respondents, when evaluating a higher income, are frequently content with more nominal income although a simultaneous rise in prices keeps real income constant. The authors conclude that preferences of people may correlate with nominal changes even if there is no real change. The attribution of well-being is driven primarily by nominal rather than a real evaluation.

Do firms anticipate this link of money illusion and fairness? The hypothesis outlined above holds that rational profit maximizing firms anticipate the link between nominal wages and workers' perception of fairness. In this case profit maximizing firms would not want to cut wages. BEWLEY (1995a, 1995b) conducted an extensive questionnaire study with employers. The author investigated why employers did not cut wages despite deteriorating economic conditions and rising unemployment rates. The author reports that businessmen are sensitive to the implications of nominal wage cuts for working morale and loyalty to the firm (see also CAMPBELL and KAMLANI, 1997; BLINDER and CHOI, 1990 or AGELL and LUNDBORG, 1995 for similar results).

C) Further areas where money illusion may be of economic relevance

This subsection briefly discusses the relevance of money illusion for various economically important areas like goods markets, portfolio selection decisions, the costs of inflation and economic policy.

Goods markets. New Keynesian theorists suggest that menu costs could be a relevant factor in explaining nominally rigid price adjustment. Motivated by this theoretical work, microeconomic evidence on price setting in individual markets has been collected (e.g. CARLTON, 1986; BLINDER, 1991, 1994). In general, price changes are found to be surprisingly infrequent. For example, CECCHETTI (1986) finds that magazines on average change their newsstand prices only every three years. KASHYAP (1995) finds that prices are adjusted only after inflation has eroded the real price by 10%. This evidence does not seem to conform with any of the (New Keynesian) theories which are based on the assumption of fully rational agents. This leads ROMER (1996: 294) to conclude: "In sum, the microeconomic evidence of price stickiness is puzzling."

The link between money illusion and fairness which seems to be potentially important in labor markets may also affect goods markets. KAHNEMAN, KNETSCH and THALER (1986) present among other findings the results from a questionnaire study in which subjects were asked to judge a store owner who increases his nominal selling price (seeing the importance of selling his goods at their current real value rather than their original nominal price). 79% of respondents found the increase in prices unfair. The authors suggest that the reason for this is that subjects are estimating profits based on nominal rather than real changes. SHAFIR and THALER (1995) run a questionnaire study to analyze whether people tend to value their possessions in terms of historic or (correctly) in terms of replacement costs. In this study, people seem to have conflicting intuitions about current value, and do not fully appreciate considerations of replacement costs. Historic cost can differ from replacement cost because of a change in the value of money or because of a change in relative prices. Thus, as SHAFIR *et al.* (1997) argue, it is not the case that money illusion can be rendered completely irrelevant by having a zero rate of inflation. CHURCHILL (1982) discusses the fact that many businesses continue to sell the old stock at old prices, despite the fact that replacement costs have gone up with inflation. Yet, it is unclear whether these entrepreneurs suffer themselves from money illusion or anticipate that customers will judge price increases as unfair.

Indexation of Contracts. Indexing is observed much less frequently than expected. Admittedly, there may be good reasons why indexation to nominal GDP will not be chosen by a fully rational agent.[30] For example, full wage indexation protects the economy from nominal shocks but exacerbates the effects of real shocks as the real wage is fixed. Even if nominal shocks were the main source of aggregate fluctuations, it remains unclear whether the wages should be indexed to the CPI, the GNP deflator, or to nominal income. However, governments frequently use unindexed contracts and have tax systems that are unindexed or incompletely indexed. FELDSTEIN (1983, 1997) attributed real effects of inflation to non-indexation of tax laws ("cold progression").[31]

A rational, risk averse decision maker is likely to prefer an indexed contract. On the other hand, a *nominally risk averse* decision maker may perceive indexed contracts as riskier since the indexed amount may end up being smaller or greater in nominal terms than a fixed dollar amount. People seem to evaluate contracts predominantly in nominal terms and are thus nominally risk averse, i.e. they prefer non-indexed (e.g. debt) contracts (see STD, 1997).

Portfolio selection. Experimental evidence on money illusion comes from a study of financial investment by THALER and TVERSKY (1996). The participants in the experiment were asked to allocate a portfolio of 100 shares between two funds. Real returns and standard deviations of the two funds corresponded approximately to the actual return of bond and stock investments. These distributions were not known to subjects, but were learned in 200 trial periods. After these trials, subjects had to make a final decision about an allocation which would be binding for 400 trial periods. The authors implemented two treatments: A non-inflation treatment and a treatment with 10% inflation. In accord with money illusion, *inflation had a profound impact on allocations*. The mean allocation to the risky (stocks) fund was 42.3% in the no-inflation treatment and 71.5% in the inflation treatment. Because of overwhelmingly positive nominal returns in the inflation condition, people exhibited much less risk aversion in that condition.

30. GORDON (1990: 1161) writes: "The reason is simple, and it is at the heart of all good microeconomics. Individual firms maximize profit by setting *their own* marginal cost equal to *their own* marginal revenue. They have no reason whatsoever to care about nominal GDP unless it provides useful information to supplement what they can learn from observing their 'local' cost and demand. There are many reasons for firms to expect their nominal marginal cost and local demand to contain idiosyncratic elements that cause them to evolve independently from nominal demand. The most straightforward argument, which is enough to make the case, is that firms in a small open economy know that their costs are determined outside the national boundaries within which domestic nominal demand applies. This principle generalizes to firms in large open economies ..." See also BLANCHARD (1990: Ch. 5). In addition, indexing is sometimes prohibited by law, as in Germany.

31. However, some countries do correct for this effect (e.g. Swiss Constitution, Art. 41[ter], §5, lit. c).

Business accounting. The widespread use of nominal accounting methods[32] may lead to substantial distortions in times of inflation. According to FISCHER and MODIGLIANI (1986: 821-2) this may "create potential misallocations of resources, partly because firm data may be misinterpreted, and partly because markets may incorrectly assess the relative desirability of investment by different firms, and provide capital at an inappropriate cost ... The use of nominal accounting methods is one example of the type of money illusion that remain in the economic system despite continuing inflation; this illusion results from the convenience of using money as a unit of account, rather than a medium of exchange function. On a priori grounds we are reluctant to believe that such illusions can remain in the system over long periods, but there does appear to be some evidence of their continued existence. They are familiar in everyday discussion; it also appears that even the supposedly sophisticated capital markets may be using nominal interest rates to capitalize real profits (MODIGLIANI and COHN, 1979). All such illusions must ultimately be self-destructive, but the surprise is that they still persist."

The cost of inflation. When investigating the costs of inflation with economic models that start from the assumption of fully rational agents, these costs appear to be relatively small compared to the large public concern with inflation as a harmful phenomenon (see SHILLER, 1997 for a survey study on why people dislike inflation). Thus, "It appears to remain the case that the man-in-the-street-notions of the costs of inflation have not been formalized in rigorous theoretical models" (DRIFFILL *et al.*, 1990: 1046).[33] Apparently, this public concern with inflation issues may actually influence economic policy. This is suggested by estimations of the impact of inflation on popularity of the governing party (see FREY, 1997 for a collection of articles).

---

32. Business accounting, methods like FIFO (First in - first out) and LIFO (Last in-first out) rely on historic prices, not replacement costs.
33. OKUN (1975) gives a broad (non-formalized) discussion of the costs of inflation. His arguments rely on the notion of "customer markets" in which the high costs of acquiring information lead to long-term relationships between workers and employers, as well as between customers and suppliers. In an economy containing both these customer markets and also "auction markets", inflation causes relative price variation, since prices in customer markets respond more slowly than those in auction markets. This, in OKUN's view, has enormous welfare costs, and he remarks: "Prolonged and intense inflation upsets many habits of economic life, confronting consumers with price increases and price dispersions that send them shopping; making them doubt their ability to maintain their living standards, and downgrade the value of their career jobs and long-term savings; and forcing them to compile more information and try to predict the future - costly and risky activities and that they are poorly qualified to execute and bound to view with anxiety." (OKUN, 1975: 383).

Chapter 4: Why can money illusion and strategic complementarity cause monetary non-neutrality?

"No one can see the spectacle in the theater or stadium if everyone stands, but who has the incentive to ... sit down?"

JAMES TOBIN (1989: 15)

This chapter explains the theoretical background of the present study. It serves the purpose of formulating the alternative hypotheses summarized in section 5.2. Section A) explains the intuition behind the theoretical argument by HALTIWANGER and WALDMAN (1985, 1989). These authors show that, when agents are heterogeneous with respect to rationality, boundedly rational agents may have a disproportionately large impact on the aggregate if strategic complementarity prevails. This intuition has come to the minds of several authors (e.g. LEIJONHUFVUD, 1981) within various contexts but will be discussed here only in the context of monetary non-neutrality. Section B) explains in what respect money illusion may be source of bounded rationality. In particular, the interaction of beliefs about others' money illusion and strategic properties is discussed.

A) Bounded rationality and monetary non-neutrality

This section provides an intuitive account of the theoretical results of HALTIWANGER and WALDMAN (henceforth HW 1985, 1989). First, we explain the concepts of strategic complements and substitutes. Second, the benchmark case of full rationality is discussed: given common knowledge of rationality (i.e. homogenous agents), anticipated money is neutral in the absence of exogenous frictions irrespective of strategic properties. In subsequent subsections, the main argument is developed. Given agents which are heterogeneous with respect to rationality, money is non-neutral. In particular, nominal rigidity and monetary non-neutrality is more pronounced with strategic complements than with strategic substitutes. This conclusion extends to the case where all agents in fact are rational, but common knowledge of rationality does not prevail.

Concepts. Strategic complementarity corresponds to the case where an increase in the strategy variable of other players will lead a profit maximizing player to increase his own strategy variable. Strategic substitutes correspond to the case where an increase in the strategy variable of other players will lead a profit maximizing player to *de*crease his own strategy (see BULOW *et al.*, 1985 for an extended discussion). The concept of strategic complementarity captures essential features of many "Keynes-type" models.[34] COOPER and HALTIWANGER (1993) provide empirical evidence on strategic complementarity in naturally occurring macroeconomies. One may therefore argue that strategic complementarity has high empirical relevance and can be thought of as a "typicial" or "natural" property of macroeconomics. In the example of monopolistic price competition, strategic complementarity corresponds to the case where an increase in the aggregate price level leads a firm to increase its own price. Strategic substitutability corresponds to the case where an increase in the aggregate price level leads the firm instead to decrease its price.[35] Put differently, strategic complementarity prevails when the best reply function *slopes upward* and strategic substitutes prevail when the reaction function *slopes downward.*

The intuition of the result by HW (1985, 1989) can be captured in a framework of monopolistic competition among firms. Suppose the economy consists of $i = 1,..., n$ symmetric agents. Let $\pi_i = \pi_i(P_i, \bar{P}_{-i}, M)$ be firm i's profit function, where P_i is the nominal price chosen by firm i and $\bar{P}_{-i}$ is the average price chosen by all other firms. The common shift parameter M denotes the quantity of money. Assume that the profit function is homogeneous of degree zero in nominal quantities, i.e.

$\lambda^k \cdot \pi_i(P_i, \bar{P}_{-i}, M) = \pi_i(\lambda P_i, \lambda \bar{P}_{-i}, \lambda M)$ with $\lambda > 0$, and $k = 0$ for all i. Further, assume that the profit function is continuously differentiable and that $\partial^2 \pi_i / \partial P_i^2 < 0$ and $\partial^2 \pi_i / \partial P_i \partial M > 0$, that is, an agent i's profit function is strictly concave in own prices, and it pays marginally to increase the nominal price after an increase of the quantity of money. Agent i's best reply for any $\bar{P}_{-i}$ has to satisfy $\partial \pi_i / \partial P_i = 0$. If the profit function has the property $\partial^2 \pi_i / \partial P_i \partial \bar{P}_{-i} > 0$, the situation exhibits strategic complementarity, if $\partial^2 \pi_i / \partial P_i \partial \bar{P}_{-i} < 0$, strategic substitutes prevail. Suppose that all agents are in equilibrium and that equilibrium is unique.[36]

---

34. See the seminal article by COOPER and JOHN (1998) or OH and WALDMAN (1994). Examples these authors give include monopolistic price competition (e.g. HART, 1982; KIYOTAKI, 1988), increasing returns of scale in production (WEIL, 1989), search in labor markets (DIAMOND, 1982, 1984; HOWITT, 1985), stock markets (SHILLER, 1984) and coordination problems (SUMMERS, 1988) to which the quotation above from TOBIN (1989: 15) refers.
35. However, this is a very unnatural case for price competition.
36. See COOPER and JOHN (1988) for the proof of the existence in the symmetric case.

Hypothesis H_0: Neutrality with common knowledge. To keep the discussion intuitive, we only discuss the case of two agents, Mrs. *i* and Mr. *j*. Suppose rationality is common knowledge, and both agents have set profit maximizing prices in an initial equilibrium. What happens when the quantity of money is halved? Both players *i* and *j* cut their nominal prices by 50% and sell the same amount of goods as before and make the same real profit. Thus, if common knowledge of rationality prevails, money is neutral *irrespective* of the strategic properties of the economic environment.

Strategic complements and heterogeneous agents. Suppose the two players are in the same initial equilibrium as before, but now agents are *heterogeneous with respect to rationality*. In particular, player *j* is assumed to be completely irrational (in the sense that *j* does not react at all to the change in the quantity of money) and *i* knows this. Accordingly, we call an agent like Mr. *j* a "naive", and an agent like Mrs. *i* a "super-rational". Naive agents are assumed to make expectation errors, super-rational agents are assumed to hold correct (i.e. rational) expectations. To form such expectations, an agent has to know the proportion of "naive" in the population, the extent to which the naive make expectation errors and how the sophisticated react on the basis of this knowledge. Since rationality requirements to form correct expectations are rather high, we will call an agent holding rational expectations a "super-rational agent".[37] What happens when the quantity of money is halved? Since Mr. *j* does not understand that he should cut his nominal price to reap the same real profit as before, *j* does not change his nominal price at all. Thus, *j*'s real price, defined as $p_j = P_j/M$ increases. If strategic complementarity prevails, Mrs. *i* will also want to increase her real price $p_i = P_i/M$ and therefore *not* cut her nominal price by 50%, but by less. Why this happens, can be illustrated in *figure 1*. Since *j* does not change his nominal price after the monetary contraction, his real price will increase by 100%. This shifts *i*'s profit function to the right. If *i* does not adjust her real price (i.e. if she cuts her nominal price by 50%), *i* will incur a loss of $\pi^*_{i0} - \bar{\pi}_{i0}$. But, since *i* is a rational, profit maximizing agent and since *i* knows that *j* is not, *i* will increase her real price to $p_i^{br}(\bar{p}_{j1})$. That is, *i* will cut her price by less than 50%. Put differently: under conditions of strategic complementarity the fully rational agent (*i*) partially *imitates* the boundedly rational agent (*j*). In this case, the *boundedly rational agents have a disproportion-*

37. Note that these agents are called "sophisticated" agents in HALTIWANGER and WALDMAN (1989).

ately large impact on the aggregate (here: the aggregate price level falls by less than 25%).

What happens to *j*'s profits in this situation? Mr. *j* will suffer a loss since he chooses the same nominal price as before which is not a best reply to *i*'s perfectly rational action.

Figure 1: Losses for an agent who does not imitate the irrational agent

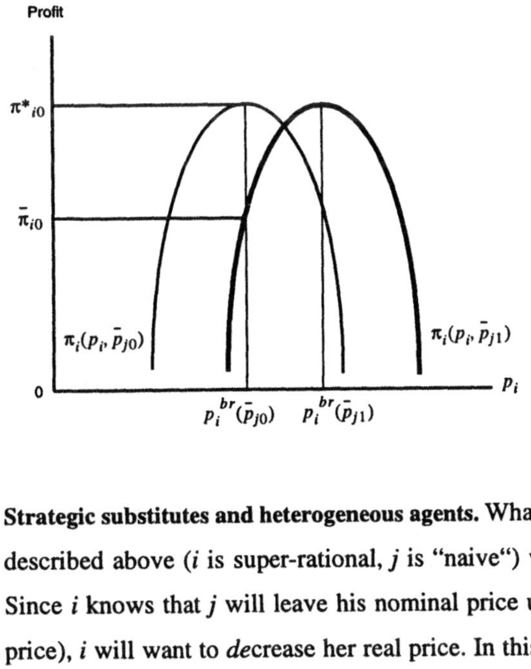

Strategic substitutes and heterogeneous agents. What happens in the situation described above (*i* is super-rational, *j* is "naive") when strategic substitutes prevail? Since *i* knows that *j* will leave his nominal price unchanged (i.e. *j* increases his real price), *i* will want to *de*crease her real price. In this case, *i*'s profit function would be shifted to left in *figure* 1. That is, *i* will cut her nominal price by *more* than 50%. Thus, under conditions of strategic substitutes, the fully rational agents partially *compensate* the behavior of the boundedly rational ones. In this case, the *boundedly rational agents have a disproportionately small impact on the aggregate outcome* (here: the aggregate price level falls by more than 25%).

Beliefs and strategic properties. Finally, suppose that both players *i* and *j* in fact are rational, but (counterfactually) *believe* that the other one is "naive". What happens after the reduction of the quantity of money in this case? Player *j* expects that *i* will not change her nominal price (which corresponds to an increase in the real price). Since *j* is a profit maximizing agent, he will (given strategic complementarity) increase his real price (i.e. cut his nominal price by less than 50%). At the same time,

i expects that j will not change his nominal price and since i is a profit maximizing agent, she will choose the price which is a best reply given her expectation. Thus, player i will also cut her price by less than 50%. As a result, both players hold *wrong expectations*, do not choose best replies to each other's actions, and, hence, incur real income losses. In this example, the aggregate price level exhibits nominal rigidity and leads to the non-neutrality of money.[38] It can theoretically be shown that the absence of common knowledge of rationality is sufficient for this conclusion to hold. Therefore, *money is non-neutral with strategic complementarity, even if none of the agents is boundedly rational, but some agents believe that other agents are boundedly rational.*

Analogy to the "guessing game". The last case discussed above can be illustrated by the "guessing game" (NAGEL, 1995), which is a special case of the well-known Keynesian "beauty contest".[39] In the guessing game, each person chooses a number from the closed interval [0,100]. The person closest to p times the average number chosen by all participants, wins a prize. For $0 \leq p < 1$, there exists a unique Nash equilibrium in which all (fully rational) players choose 0. Suppose you participate in this game. Suppose you are a rational player and therefore you know that the Nash equilibrium is zero. What number would you choose, if you believe that some players are not rational? Since the Nash equilibrium is on the boundary of the support, any deviation from equilibrium must be in one direction. Hence, boundedly rational players will choose strictly positive numbers. Strategic complementarity prevails in this game because the best reply (i.e. the winning number which is p times the average) increases with the average number chosen. Thus, as a rational person, the number you will want to choose is the higher the stronger your belief that others are irrational. Since the guess-

38. With strategic substitutes, nominal prices would overshoot and money would be non-neutral.
39. KEYNES (1936: 156) speculated that the market value of assets in financial markets may become unhinged of "fundamentals" and become dependent on free-floating expectations: "... professional investment may be linkened to those newspaper competitions in which the competitors have to pick out the six prettiest faces from a hundred photographs, the prize being awarded to the competitor whose choice most nearly corresponds to the average preferences of the competitors as a whole; so that each competitor has to pick, not those faces which he himself finds the prettiest, but those which he thinks likeliest to catch the fancy of the other competitors, all of whom are looking at the problem from the same point of view. It is not the case of choosing those which, to the best of one's judgment, are really the prettiest, nor even those which the average opinion genuinely thinks the prettiest. We have reached the third degree where we devote our intelligences to anticipating what average opinion expects the average opinion to be. And there are some, I believe, who practice the fourth, fifth and higher degrees."
For experimental investigations of similar games (coordination games that have many equilibria) see VAN HUYCK *et al.* (1990).

ing game can be solved by iterated elimination of dominated strategies, you may, for example, try to guess how many iterations other players are able to perform to get an estimate of the average number chosen.[40] Alternatively, you may think that players who are not fully rational, randomly choose a number between 0 and 100. In this case, you may, as a rational person, choose a number like p times 50. In any case, you will choose a number which is a best reply to the average number you expect to be chosen. Your expectation of the average number depends on the degree of rationality you ascribe to other players. Or, on your belief on the belief of other players of the rationality of the rest of the players etc. (see footnote 39).

Hypothesis H_{A2}. The hypothesis (which will be explained in more detail below) derived from the argument by HW is the following: Strategic properties matter because super-rational agents react differently to the behavior of "naive agents". If strategic complements prevail, naive agents cause disproportionate nominal rigidity and monetary non-neutrality because the super-rational agents tend to imitate them. If strategic substitutes prevail, naive agents cause disproportionately small nominal rigidity and monetary non-neutrality because the super-rational agents tend to compensate their effect. If the behavior of "naive agents" does not have anything to do with money illusion, the effects of strategic properties should be the same irrespective of nominal vs. real representation. That is, super-rational agents should hold the same (correct) expectations about aggregate behavior irrespective of representation.

B) Money illusion and strategic properties

In the framework explained above, money is neutral if common knowledge of rationality prevails (see H_0 on page 38). In contrast, if all agents are naive, money will be massively non-neutral. In section A), we assumed that a given portion of agents may be irrational ("naive") for *some* reason. We now discuss how money illusion as a source of irrational behavior enters the picture. First, money illusion may directly cause aggregate effects because it affects the proportion of naive agents and their behavior. Second, money illusion may indirectly cause aggregate effects because some agents strategically react to other agents' money illusion. beliefs about other agents' behavior. Two hypotheses on how money illusion affects aggregate behavior are formulated.

---

40. See also HO, CAMERER and WEIGELT (1998).

Effects of first-order money illusion according to H_{A1}. Chapter 3 explained that money illusion is induced by the nominal vs. real representation of the economic environment. One may hypothesize that subjects are led to take suboptimal actions because of representation. These agents will be called prone to "first-order money illusion". If this is the main reason of monetary non-neutrality, and if the proportion of agents being prone to first-order money illusion increases with the number of nominal elements in representation, one arrives at the simple money illusion hypothesis H_{A1}. This hypothesis bluntly states that nominal vs. real representation matters because more people are prone to first-order money illusion in a nominal representation. Note that H_{A1} does not take into account the main proposition of section A), i.e. that strategic properties matter because they provide super-rational agents with different incentives for given disequilibrium expectations (see H_{A2} above). To illustrate, consider first the simplest case where agent *j* acts in complete isolation, i.e. is not in a strategic situation. In this case, his decision can be affected exclusively by "first-order money illusion". That is, Mr. *j* may choose non-optimal actions for two reasons: First, he may incorrectly perceive the real economic decision environment because he confuses real and nominal magnitudes. This is illustrated by arrow *a* of *figure* 2. Second, Mr. *j* may actually prefer certain nominal values over others. This is illustrated by the dotted arrow *b*. If strategic behavior is unimportant, first-order money illusion directly translates to the aggregate level. However, as is explained next, strategic properties may affect money illusion.

Figure 2: Effects of first-order money illusion according to H_{A1}

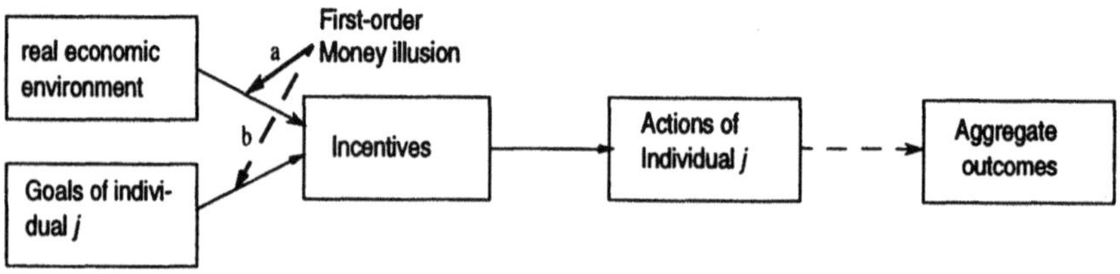

Fisher effect, high-number illusion and "naive agents". HALTIWANGER and WALDMAN (henceforth HW) leave open the question why naive agents may not respond to a change in economic incentives. Money illusion may explain why naive agents do not respond to a negative[41] nominal shock. As explained in chapter 2, we distinguish between two aspects of representation inducing money illusion: The FISHER effect and the high-number illusion. The FISHER effect (or, nominal anchoring) would in general predict an inertial reaction of naive agents, no matter which strategic properties prevail. The effect of high-number illusion on the behavior of naive agents, on the other hand, depends on strategic properties. Given strategic complementarity and a monetary contraction, naive agents would tend to reduce nominal prices only a little. The reason for underadjustment is as follows: For a given price level, the highest (nominal and real) payoffs by definition prevail along the best reply. For given real payoffs, nominal payoffs tend to be the higher the higher the price level. Since the best reply has a positive slope with strategic complements, high nominal payoffs prevail at *high individual* prices. Suppose naive subjects are prone to high-number illusion, i.e. suppose high nominal payoffs that go with high nominal price levels are taken for high real payoffs. In this case, a naive agent would choose high nominal prices. In conjunction with a negative monetary shock, this would mean that naive subjects make sticky price choices. Things are different, however, with strategic substitutes. In this case, the best reply function has a negative slope. Hence, high nominal payoffs prevail at *low individual* prices. An agent prone to high-number illusion would hence, tend to choose low prices after a negative monetary shock. To summarize: First-order money illusion prevails if agents are confused by the nominal representation and take, as a consequence, non-optimal actions. First-order money illusion provides an explanation for the behavior of naive agents. There are two aspects of nominal representation inducing money illusion. If nominal anchoring is the only important aspect, money illusion leads to inertial behavior of naive agents irrespective of strategic properties. If, on the other hand, high-number illusion is the only important aspect of money illusion, behavior of naive agents depends on strategic properties. It leads to inertial behavior if strategic complements prevail, but to a strong (over-)adjustment of prices if strategic substitutes prevail.

Aggregate level effects of first-order money illusion. Section A) explained that a rational person will choose a price on the basis of the expected average price. This expectation in turn depends on the presumed effect of money illusion on other players' price choices. To illustrate, consider two agents *i* and *j* which are in a strategic situa-

41. In the experimental study explained in Part II, we implement a negative monetary shock. The reasons for this choice will be explained in Part II (see in particular the discussion of *figure* 8 on page 74).

tion. Suppose Mrs. *i* is super-rational, i.e. Mrs. *i* is free of first-order money illusion *and she knows* how Mr. *j*'s behavior is affected by first-order money illusion. Thus, Mrs. *i* has perfect foresight and correctly anticipates Mr. *j*'s action. If the decision environment is characterized by strategic complements (substitutes), *i* will want to increase (decrease) her real price if she knows that *j* increases his because of money illusion. Thus, if the behavior of naive agents is independent of strategic properties, the super-rational agent in both conditions holds the same and correct expectation about the behavior of first-order money illusion prone Mr. *i*, but reacts differently to this same expectation. As a consequence, first-order money illusion prone Mr. *j* has a disproportionately large impact on the aggregate when strategic complements prevail, and a disproportionately small impact on the aggregate when strategic substitutes prevail. The important thing to note is that the aggregate effect of *given* first-order money illusion depends on strategic properties. Put differently, (given first-order) money illusion matters more with strategic complements than with substitutes. This is illustrated in *figure* 3. The lower half of this figure adds the super-rational player *i* to *figure* 2. This player is assumed to hold correct expectations on the behavior of the naive player (see arrow *f*) and, since she acts rationally given her expectation, will choose different actions in different strategic environments (see arrow *e*). The naive player *j* is assumed to hold wrong expectations because he is prone to money illusion himself (see arrow *c*).

Aggregate effects of higher-order money illusion. Finally, suppose that *i* is not a super-rational, but only a "sophisticated" agent. By this we mean that agent *i* is free of first-order money illusion and anticipates that Mr. *j* may suffer from first-order money illusion, but she *does not have perfect foresight* about the behavior of Mr. *j*. Thus, *i*'s action depends on her belief about how *j*'s action will be affected by money illusion. Higher-order money illusion is said to prevail if expectations of agents which are themselves free of first-order money illusion are systematically affected by the nominal vs. real representation. Hence, nominal vs. real representation may affect expectations, and these expectations (not the first-order money illusion itself) may cause money illusion to matter in the aggregate.

Figure 3: Aggregate effects of first-order money illusion on behavior according to H_{A3}

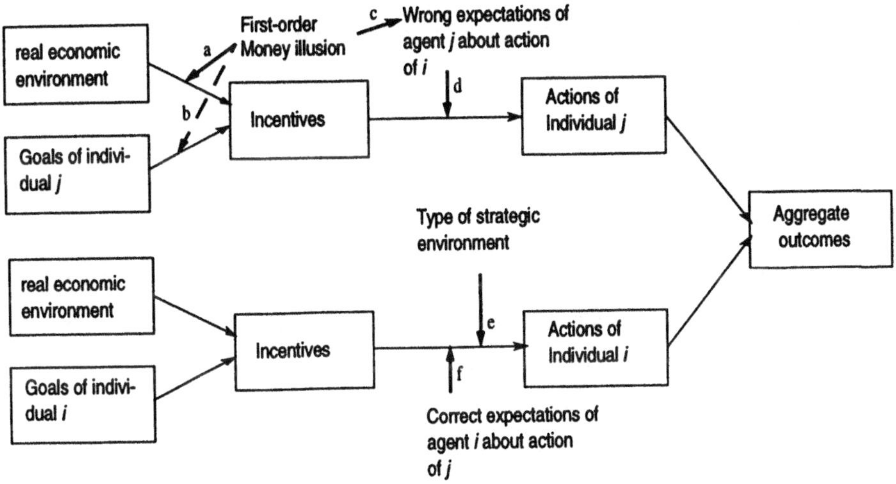

Hypothesis H_{A3}. Hypothesis H_{A3} combines the hypotheses H_{A1} and H_{A2}. According to H_{A3}, strategic properties are crucial for how much first-order money illusion matters for aggregate behavior. The hypothesis H_{A3} states that "money illusion matters more with strategic complements", and "strategic properties matter more when money illusion is prevalent". The first part of the hypothesis results from direct and indirect effects. First, as suggested by H_{A1}, direct effects will be more pronounced with complements because both aspects of money illusion draw the behavior of first-order money illusion prone subjects away from equilibrium if complements prevail, but not if substitutes prevail. Second, as suggested by H_{A2}, indirect effects of money illusion will be more pronounced with complements because sophisticated agents react differently to the behavior of the naive. A given degree of first-order money illusion is magnified when strategic complements prevail, but mitigated when substitutes prevail. The second part of the hypothesis ("strategic properties matter more when money illusion is prevalent") is deduced from the following argument: If there are no naive agents, strategic properties do not matter (as suggested by H_0). If money illusion explains the (supposed or actual) behavior of naive agents, strategic properties matter the more, the more naive agents there are.

Chapter 5: Summary of part I and hypotheses

Section 5.1. summarizes Part I and section 5.2 summarizes the hypotheses of the experimental study.

5.1. Summary

The question of whether, and if so, why money affects output is of paramount importance to economic theory and policy. It is important to theory because this question is at the center of how the real and the monetary side of the economy are interconnected; it is important for economic policy since the "inappropriate management of the money supply ... may result in serious damage to the economy" (EATWELL *et al.* 1989: xi). These authors explain that there are two conflicting intuitions about money. The intuition that money should be neutral arises from basic (microeconomic) theory, the intuition that money should be non-neutral from (macroeconomic) empirical evidence. Yet, as pointed out by LUCAS (1996: 661) "the question has not been given anything like a fully satisfactory answer".

Chapter 1 has shown that the empirical "evidence concerning whether money shocks have important real effects is controversial" (ROMER, 1996: 241). The issue remains disputed because there may be substantial mismeasurement in macroeconomic aggregates, and the inferences drawn from empirical research on monetary neutrality prove to be sensitive with respect to measurement errors. Moreover, it seems to be impossible to isolate truly causal relations from empirical field data.

Chapter 2 has shown that economic theorists have chosen very different approaches to explain monetary non-neutrality. Researchers of the Real Business Cycle tradition do not seem to think that monetary shocks are of any importance to explain real aggregate fluctuations. New Classical economists argue that only unanticipated changes in money affect real economic activity whereas anticipated changes in money should be neutral in a friction-free environment according. New Keynesians introduce exogenous nominal frictions into incompletely competitive economies to explain monetary non-neutrality. Despite the vast differences in approaches between these lines of macroeconomic thought, all of them start from the assumption of fully rational agents holding rational expectations. In contrast to these approaches, some theorists have recently started to investigate the implications of relaxing the *assumption* of full rationality of economic agents. A general finding from this literature is that a small relaxation of this assumption may have substantial consequences at the aggregate level.

Chapter 3 discussed money illusion as the most important deviation from full rationality in the context of monetary non-neutrality. Economic psychology explains that money illusion arises because of nominal anchoring and framing effects. Anchoring prevails when people tend to judge changes in the economic environment from some naturally given (by experience or custom) starting point. Framing prevails when alternative (nominal) representations of the same (real) economic environment systematically affect individual decisions. Two aspects of nominal representation may affect behavior. According to IRVING FISHER (1928), people may fail to perceive that a constant smallest nominal accounting unit expands or shrinks in value when the quantity of money is changed (FISHER effect). Second, people may tend to generally interpret high nominal incomes as high real incomes because they fail to correctly deflate nominal values. In this case, people are said to be prone to high-number illusion. A questionnaire study by SHAFIR, TVERSKY and DIAMOND (1997) shows that money illusion is a pervasive phenomenon at the individual level. In particular, it seems to be the case that people expect other people to be prone to money illusion. Hence, money illusion could potentially be relevant to understand many aspects of the economic life.

Chapter 4 discussed how money illusion and strategic complementarity may cause money to be non-neutral. Strategic complements and nominal representation is the "natural" case, i.e. these conditions are typical for many macroeconomic relations. Under these conditions it is extremely restrictive to dismiss money illusion as an explanation of monetary non-neutrality. The work of HALTIWANGER and WALDMAN (1985, 1989) suggests that super-rational agents react differently to the presence of naive agents in different strategic environments. If the economic environment is characterized by strategic complements, super-rational agents tend to imitate the behavior of naive agents; if it is characterized by strategic substitutes, super-rational agents tend to compensate the behavior of the naive. Money illusion may explain why naive agents do not react to the monetary shock. This holds in particular if strategic complements prevail. In this case, both the FISHER effect and the high-number illusion draw behavior away from equilibrium. First-order money illusion may directly produce aggregate effects because more people fail to take optimal decisions. In the presence of strategically anticipating agents, these effects are aggravated with complements, and mitigated with substitutes.

5.2. Hypotheses

Along the lines of the discussion in chapter 4, the hypotheses with respect to monetary non-neutrality are summarized below. The New Classical hypothesis is the Null-hypothesis against which several alternative hypotheses are tested. The Null hypothesis states that an anticipated monetary shock should be neutral in the absence of nominal frictions. The implicit assumption behind this proposition is that common knowledge of rationality prevails. Thus, if this implicit assumption holds, and all agents are perfectly informed about the change in money, and if prices are (exogenously) free to adjust, money should in fact be neutral. This leads us to formulate the following hypothesis, which will also be called "Neutrality"- hypothesis:

H_0: *An anticipated monetary shock is neutral in the absence of exogenous frictions*
(irrespective of the representation of the environment and
irrespective of the strategic properties of the economic environment).

The research in economic psychology discussed in chapter 3 provides suggestive evidence that money illusion is a pervasive phenomenon at the individual level. Economic psychology explains money illusion to arise from representation effects. From these findings, the general hypothesis is drawn that "nominal vs. real representation matters". Thus, representation makes it more or less difficult to "pierce the veil of money" and, consequently, non-neutrality after an anticipated shock should be more pronounced when there are more nominal elements in representation. This hypothesis does therefore not account for strategic effects. The following hypothesis, which will also be termed (non-interactive or simple) "Money illusion" hypothesis is formulated:

H_{A1}: *The more nominal elements there are in the representation of the economic environment, the larger the extent of monetary non-neutrality*
(irrespective of the strategic properties of the economic environment).

Recent developments in economic theory, especially by HALTIWANGER and WALDMAN (1985, 1989), stress the importance of the strategic properties of the economic environment, if agents are heterogeneous with respect to rationality. According to this theory, the strategic properties of the economic environment matter because (super-)rational agents (correctly) expect "naive" agents not to respond to a change in money supply and react differently to this non-response in different strategic situations. From this line of theorizing, the following hypothesis, which will also be called the "HALTIWANGER / WALDMAN"- hypothesis, is deduced:

H_{A2}: *When the economic environment is characterized by strategic complementarity, non-neutrality after an anticipated monetary shock is more pronounced than with strategic substitutes*
(irrespective of the representation).

Hypothesis H_{A3} combines the hypotheses discussed above and suggests that money illusion and strategic properties interact. If there are no naive agents, money will be neutral irrespective of strategic properties (see H_0). Money illusion may provide an explanation for the incidence of naive agents. Hence, aggregate behavior can be expected to differ more with different strategic properties when money illusion is prevalent. However, the aggregate effects of money illusion also depend on strategic properties. There are two aspects of representation which induce money illusion, i.e. anchoring and high-number illusion. These two aspects may affect the behavior of the naive differently in the two strategic conditions. Nominal anchoring leads to an inertial reaction of the naive irrespective of strategic properties. If strategic complements prevail, super-rational agents would respond to this by choosing sticky prices, and sticky aggregate prices would result. If strategic substitutes prevail, super-rational agents would respond by adjusting prices a lot, and the aggregate price level would thus be relatively close to the equilibrium value. Things are more complicated with high-number illusion: First-order high-number illusion draws behavior of the naive after a monetary contraction "far away" from equilibrium with complements, but draws their behavior "relatively close" to equilibrium with substitutes. Hence, the direct impact of money illusion is larger (smaller) when strategic complements (substitutes) prevail and this larger (smaller) effect is even magnified (mitigated) by anticipating agents.

The hypothesis H_{A3} will also be labeled as "Money illusion *cum* strategic anticipation"- hypothesis:

H_{A3}: *Strategic properties matter more* (in the sense of H_{A2}) *when there are many nominal elements in the representation of the economic environment, and representation matters more* (in the sense of H_{A1}) *with strategic complements than with strategic substitutes.*

It should be noted that the "Neutrality"- hypothesis is a "strong" hypothesis in the sense that it makes a precise point prediction, i.e. that anticipated money should be exactly neutral, whereas the other hypotheses are qualitative hypotheses of the style "the more of the one, the larger the other".

Part II presents a design to test these hypotheses[42] and to isolate the causes of monetary non-neutrality after an anticipated shock.

42. The hypotheses will be operationalized in more detail with respect to expectations formation on the basis of the findings discussed in chapter 1. See Part III, sections 2.1. and 3.1.

Part II

Experimental Study

Chapter 1 briefly explains important aspects of the experimental method. It is argued that the application of experimental methods may provide useful insights when investigating monetary non-neutrality. In particular, macroeconomics should be viewed as an indirectly experimental science.

Chapter 2 describes the experimental design to isolate causal factors of monetary non-neutrality.

Chapter 1: Are experiments in macroeconomics possible?

"Economists ... cannot perform the controlled experiments of chemists or biologists because they cannot easily control other important factors. Like astronomers or meteorologists, they generally must be content to observe."

PAUL A. SAMUELSON and WILLIAM D. NORDHAUS (1985: 8)

This chapter discusses whether experimentation in economics, and in particular in macroeconomics is possible and useful. Sections A) and B) provide a general introduction which draws heavily on FALK and TYRAN (1997). Section C) explains that macroeconomics should be viewed as an indirectly experimental science and briefly summarizes the small existing experimental literature on (monetary) macroeconomics.

A) What is experimental economics?

Experimental economics is still a relatively young discipline and has become an accepted method of economics only in the past few decades. Basically, economic experiments consist in observing the behavior of real people who are motivated by economic incentives in a controlled environment. Experiments and theory complement each other: On the one hand, theory provides the background to develop an experimental design. On the other hand, the experimental method is an important stimulus to economic theory. Systematic refutation of some theories may motivate economic theorists to search for behaviorally more relevant theories (see, e.g., FEHR and SCHMIDT, 1998) which may then be experimentally tested in turn.

The quotation above illustrates that not even two decades ago eminent economists were very sceptical about the viability of controlled experiments in economics.[1] Yet, a similar scepticism prevailed for long periods in other sciences where experimental methods are accepted today. For example, physics was from the times of ARISTOTELES (384 - 322 B.C) to the times of FRANCIS BACON (1561 - 1626) and GALILEO GALILEI (1564 - 1624) thought be a non-experimental science. Similarly, biology was thought to be non-experimental before the work of MENDEL in the 19th century, likewise psychology until the 20th century (see FRIEDMAN and SUNDER, 1994: 121-32).

1. This is even more the case with public perception. For example the Encyclopedia Britannica (1991: 395) states that "there is no laboratory in which economists can test their hypotheses."

Today, experimental economics is an important method in various areas of economic research, such as game theory, industrial economics, finance, or public choice (see KAGEL and ROTH, 1995 for surveys of experimental research). In view of this successful development it is not surprising that SAMUELSON and NORDHAUS have erased the above citation from their textbook. Instead, they describe experimental economics (1992, 14th.ed.: 5) as an "exciting new development".

B) What is an experimental design?

Chapter 2 of this Part describes in detail the experimental design of this study. In this section, some basic aspects of an experimental design are discussed. Economic theories are, just as economic experiments essentially characterized by three elements (for the terminology, see SMITH, 1976, 1982, 1989): the environment, the institution, and the (predicted and observed) behavior. Economic theory predicts behavior on the basis of the interaction of the environment and the institution. This theoretically predicted behavior can be compared to the actually observed behavior in the experiment. To be able to compare the two, the experimental design has to implement the essential characteristics of the environment and the institution, i.e. the theory.

Environment. The environment consists of a number of agents with certain characteristics. These characteristics are e.g. captured by a preference structure, resource endowments, cost functions, and the like. In theory, agents are assumed to have these characteristics. But in an experiment, these characteristics have to be implemented or "induced" to the agents. In the design of the present study (see chapter 2), the environment consists of the number of agents in the economy and their payoff function. A necessary precondition for a successful experiment is that experimental subjects act according to the preferences which are induced by the experimentalist. To achieve this end, several conditions have to be met. For example, there should be non-satiation in the medium of payment (usually money), and payoffs should be dominant over other costs or benefits that accrue to an experimental subject in the course of the experiment. If, for example, subjects are bored or tired because the experiment lasts too long, experimenters lose control over the behavior. Usually, dominance is thought to be achieved, if experimental subjects earn their opportunity cost. For undergraduate students, this would correspond to the locally paid wage for simple jobs. The design presented in the next chapter has been chosen to investigate the effects of controlled variations in the environment on the aggregate outcome.

Institution. As explained above, a theory makes assumptions not only about the environment, but also about the institution within which economic agents interact. An institution can be thought of as a set of rules delimiting the set of individual actions. The institution determines for example, who can do what, when and for how long. The behavioral properties of institutions can be examined by controlled variations of the institution. A prominent example is the comparison of different market institutions with respect to the speed of convergence to equilibrium and overall efficiency (see DAVIS and HOLT, 1993 for an introduction).

Behavior. Given the environment and the institution a theory makes a prediction. For example: "If all agents are rational profit maximizers, well-informed about the economic environment, and there are no exogenous nominal frictions in the market, then money is neutral". Experiments allow to test such a prediction. If the environment and the institution is implemented in the way theory assumes, observed behavior can be compared with the theoretical prediction. If observed behavior systematically deviates from the predicted behavior, one should be very sceptical that this theory will work in more complex environments.

If a theory does not prove to be successful in predicting observed behavior, the *causes* of this failure can be investigated by means of experimental economics. One possibility (the one chosen in the present study) is to vary the environment in a controlled *ceteris paribus* fashion to identify conditions under which the theory is more successful. A ceteris paribus variation is defined as change of only one aspect of the environment. Comparison of behavior in the respective treatments allows to isolate causal factors. If the cause of the failure to predict actual behavior is isolated, theoretical work receives new impulses. Theorists are then called upon to incorporate the experimental finding into their models. The opposite is also possible: When competing theories make differing predictions under same conditions (environment and institution), experimental methods allow to discriminate between theories, i.e. to test which theory has more predictive power in which environment.

C) Macroeconomics as an indirectly experimental science

"Perhaps macroeconomics too, like meteorology and astronomy, will become an indirectly experimental discipline, one that relies on experimentally verified results in constructing its central theories, although the central theories themselves are not amenable to direct experimental examination."

DANIEL FRIEDMAN and SHYAM SUNDER (1994: 2)

Many economists still seem to believe that it is impossible to implement macroeconomic experiments. This opinion may be due to the fact that there are different notions of what exactly constitutes a macroeconomic experiment. Regime shifts in politics are sometimes called "macroeconomic experiments" (see e.g. BUITER and MILLER (1981) on the "THATCHER experiment"). However, such regime shifts do not constitute controlled experiments in the sense discussed above.

This section briefly reviews the literature on macroeconomic laboratory experiments. As discussed above, laboratory economics is concerned with observing the behavior of real people who are motivated by economic incentives in a controlled environment. We will hence not discuss simulations which are sometimes called "computational experiments" since they do not observe the behavior of real people (see e.g. KYDLAND and PRESCOTT, 1996). Section a) briefly reviews existing approaches in experimental macroeconomics. Section b) focuses on experiments addressing issues in monetary macroeconomics.

a) Experimental macroeconomics

Given the rapid expansion of the experimental economics literature, experiments concerned with macroeconomic issues are still relatively rare. WOLTJER (1996) provides a survey on experimental macroeconomics. He argues that the main reason for the neglect of macroeconomic issues in experimental economics is to be found in the complexity of these issues. Some researchers have tried to capture this complexity and implemented environments where subjects are given hundreds of informations, and have to decide on dozens of variables. In such environments, subjects almost inevitably get confused and data may become very difficult to interpret. In these circumstances it is almost impossible to isolate causal mechanisms because too many variables and conditions change simultaneously. Other researchers viewed macroeconomics as an indirectly experimental science. These researchers experimentally test the assumptions underlying macroeconomic theories, even though they may not be able to test the macroeconomic theories themselves (see quotation above). To be able to test assumptions underlying macroeconomic theories, these researchers have been eager to study environments which are as simple as possible, thus abstracting of many interdependencies which may be important in actual macroeconomies.

Complex macroeconomic environments. An example of how difficult experimental data become to interpret is STERMAN (1989) who investigates one aspect yielding complexity: intertemporal macroeconomic relations. He implements a relatively simple multiplier-accelerator economy. Subjects had to manage the investment goods industry of the economy. At the beginning of the experiment the economy is in steady state equilibrium but quickly (endogenously) exhibits wild fluctuations. WOLTJER (1996) criticizes the experiment which in his view is "not about macroeconomics but about human capabilities for solving a dynamic stock management problem." (See also FIEDLER, 1979 for a similar case).

TIETZ (1972, 1996) uses an even more complex macroeconomic experimental game called KRESKO to investigate collective bargaining and central bank decision making. The KRESKO game implements a monetary economy with a rudimentary state sector and four other sectors (households, firms, banks, and central bank) which are connected by five markets. The model is a system of nonlinear differential equations with regime switches. Subjects take the roles of labor or employers' union leaders who bargain over labor market conditions. There is a third person in the role of the central bank which decides over monetary policy parameters. Subjects receive about 200 informations on macroeconomic variables in each period. Hence, subjects are in a complex bargaining situation where the environment of the bargaining procedure is taken from a macroeconomic system.

Simple environments: Testing assumptions underlying macroeconomic theories. There is some relatively recent experimental work which aims at testing assumptions or elements of theories which are central to more complex macroeconomic theories.

Experimental economists have only very recently begun to tests theories which claim to explain "sticky" prices. WILSON (1998) is the first paper to experimentally investigate the "menu cost" explanation for sticky prices. He finds that menu costs may explain sticky price adjustment in monopolistic price setting in static (but not in dynamic) environments. CASON and FRIEDMAN (1998) investigate reasons for price stickiness after cost shocks in markets where consumers have to bear search costs (customer markets). They find that prices fluctuate less than costs in such markets.

Experimental research on (real) wage rigidities as explanations for involuntary unemployment has been initiated by FEHR, KIRCHSTEIGER and RIEDL (1993). These authors test the gift exchange type efficiency wage models (AKERLOF, 1982). They find that reciprocity in the guise of gift exchange is a robust phenomenon (see also FEHR and FALK, 1998). HEY and DICAGNO (1998) investigate the "dual decision hypothesis" of CLOWER which was a popular explanation of non-market clearing and hence the basis of a theory of involuntary unemployment. These authors find that the competitive equilibrium outcome is not obtained with sequentially opening markets.

b) *Experiments in monetary macroeconomics*

DUFFY (1998) provides a survey on monetary economics in the laboratory. His primary focus is on the different roles and types of money in the economy.

Inflation in OLG models. Perhaps the best-known experiments in macroeconomics are those on inflation and business cycles using overlapping generations (OLG) models (see OCHS, 1995 or SARGENT, 1993 for surveys). They focus on equilibrium selection in an OLG framework. For example, MARIMON and SUNDER (1993) assume that people live two periods in which they try to maximize lifetime utility that depends on consumption in the two periods they live. Saving is the main decision variable. The members of the young generation save, they exchange money for consumption goods. The value of money in the current period depends on the exchange rate between money and goods in the next period, i.e. the inflation rate. There are two equilibrium inflation rates in this model. The pareto-optimal inflation rate is relatively low and stable with adaptive expectations, while the high inflation rate equilibrium is stable with rational expectations. Experimental results show that there is a tendency towards the adaptive expectations equilibrium.[2] In contrast to those studies, KIRCHKAMP and BERNASCONI (1998) find in a similar setting[3] that monetary policy may strongly affect saving decisions.[4]

---

2. Since expectations formation is a crucial element of macroeconomics, one could also count all experimental studies on expectations formation as being relevant for macroeconomics. However, we do not attempt to survey this broad literature (see e.g. DANIELS and PLOTT, 1988).
3. The differences are mainly framing differences. For example, subjects describe their expectations graphically and can test the implication of several different expectations before making a saving decision. The market is presented to subjects as a market operating in the EMU and monetary policies have labels.

Exchange rates. ARIFOVIC (1993) designed an experiment to investigate exchange rate behavior. Subjects' key decision variable is saving. In addition to choose the amount to be saved, subjects had to decide on the allocation of their savings on the two currencies. The experimental results are difficult to interpret and could not be explained by either adaptive nor rational expectations. NOUSSAIR, PLOTT AND RIETZMAN (1998) conduct multiple unit double auction markets (MUDA) to analyze exchange rate formation and other questions of international economics. They provide some support and some important qualifications for the basic assumptions in this field (e.g. purchasing power parity, law of comparative advantage).

Pro's and Con's of experimental macroeconomics. Macroeconomic experiments are necessarily much simpler than the "real world", but some studies are much more complex than some of the very stylized general equilibrium theories. Hence, one may view experiments as an effort to bridge the gap between theory and the complex historical ("happenstance") data generated by the "real world" (e.g. TIETZ, 1996). There is an inherent problem of the indirectly experimental approach to macroeconomics. The question is how far can and should one go by abstracting? A basic trade-off arises between external validity of the results (also called "parallelism to the naturally occurring economy") and internal validity ("control"). If the experiment is too complex, the data may be hard to interpret and the experiment may altogether not be very telling.[5] If the experiment is too simple, one should be very careful to infer policy implications from experimental results (see Part IV for an extended discussion). In our view, the useful approach is to test fundamental assumptions on which macroeconomic theories are built. If some of these fundamental assumptions prove to behaviorally wrong, one should be very sceptical that such a macroeconomic theory serves as a useful guide for economic policy.

---

4. For somewhat different approaches see also LANGDANA (1994) who tests the Lucas 'islands' model, as well as LIAN and PLOTT (1994) who conduct general equilibrium experiments in an environment with multiple unit double auction markets with fiat money and bonds.
5. However, running complex experiments and simulation studies may be very useful for explorative purposes.

Chapter 2: Experimental design to isolate causes of monetary non-neutrality

"Experimental models can allow you to see, on a small scale, the essence of events that occur in the vastly larger and more complex real world. ... but how could you experimentally produce recessions and recoveries?"

PAUL KRUGMAN (1994: 29)

This chapter presents an experimental design which allows to implement an anticipated monetary shock and to observe nominal as well as real variables (and, thus, monetary (non-)neutrality). The discussion proceeds as follows: Section 2.1. provides a general description of the design. Section 2.2. explains procedures and parameters. Section 2.3. describes the treatments in detail. Section 2.4. mentions advantages of the experimental design in the investigation of monetary non-neutrality.

2.1. General description of the experimental design

To study under which circumstances a monetary shock is (non-)neutral an n-player pricing game was implemented, i.e. experimental subjects (acting as firms) simultaneously chose nominal prices in consecutive periods. The pricing game was divided into a pre- and a post-shock phase. At the beginning of the post-shock phase, an exogenous and anticipated nominal shock was implemented. In total, there were five treatments which implemented *ceteris paribus* variations in two dimensions. Each treatment condition had 40 periods. During the first 20 periods of a session the money supply was given by M_0. Then a fully anticipated monetary shock was implemented by reducing the money supply to M_1. This shock and the fact that the post-shock phase lasted again 20 periods was common knowledge. Our major interest concerns subjects' pricing behavior and the associated real effects in the post-shock phase. The pre-shock phase mainly serves to make the subjects acquainted with the computer terminal and the decision environment. In addition, and more importantly, the pre-shock phase allows to see whether subjects reach the equilibrium in each of the treatment conditions.

Each subject of an experimental session belonged to a group of n players. The group composition remained unchanged for all 40 periods. The real payoff for agent i was given by

$$\pi_i(P_i, \bar{P}_{-i}, M), \qquad i = 1, ..., n \tag{1}$$

where P_i denotes *i's* nominal price, $\bar{P}_{-i}$ represents the average price of the *other* n-1 group members while M denotes a nominal shock variable (money supply). The payoff functions (1) have the following properties:

(*i*) They are homogeneous of degree zero in P_i, $\bar{P}_{-i}$ and M.

(*ii*) There is a unique best reply for any $\bar{P}_{-i}$.

(*iii*) The best reply is (weakly) increasing (strategic complements) or decreasing (strategic substitutes) in $\bar{P}_{-i}$.

In addition the functional specification[6] of (1) implies that the Nash equilibrium

(*iv*) is unique for every M,

(*v*) is the only Pareto efficient point in payoff space, and

(*vi*) can be found by iterated elimination of weakly dominated strategies.

Note that π_i does not depend on the average price $\bar{P}$ of all group members but on $\bar{P}_{-i}$. This feature makes it particularly easy to play best reply for a given expectation about the other players' average price. If π_i were made dependent on $\bar{P}$, so that P_i affects $\bar{P}$, it would have been much more difficult for i to compute the best reply.

Property (*i*) was implemented because the analysis focuses on the adjustment process of an economy with money-neutral (real) equilibria. To see that property (*i*) implies neutrality note that a change in M from M_0 to $\lambda M_0 = M_1$ leaves real payoffs unaffected if prices change to λP_i and $\lambda \bar{P}_{-i}$. Moreover, if P_i, $i = 1, ..., n$, is a best reply to $\bar{P}_{-i}$ at M_0, λP_i also is a best reply to $\lambda \bar{P}_{-i}$ at λM_0. Thus, λP_i^* for all i is the post-shock equilibrium.

Property (*ii*) was chosen because it is likely to speed up adjustment to the equilibrium. At the end of each period each player was informed about the realization of $\bar{P}_{-i}$. Since i knew that all the other players had unique best replies the realization of $\bar{P}_{-i}$ was more informative.[7] Property (*iii*) implements strategic complements or strategic substitutes and was implemented for the reasons given in Part I, chapter 4. In princi-

---

6. The functional form is presented in appendix A3.
7. If the other n -1 players have multiple best replies and choose more or less randomly among them, $\bar{P}_{-i}$ exhibits (*ceteris paribus*) more randomness compared to a situation with unique best replies. Note that due to the discretization of price choices multiple best replies are present in the pre-shock phase of some treatments. However, around the equilibrium unique best replies always prevail.

ple, a monetary shock can have a permanent real effect on an economy with multiple equilibria, if – by affecting disequilibrium dynamics – it has an impact on equilibrium selection. Yet, since it is better to study the simple problems before the more difficult ones' a unique equilibrium is implemented (property (*iv*)). Property (*v*) is very untypical for a pricing game. However, important (experimental) reasons suggested to implement property (*v*). In particular, the property was implemented to rule out that both money illusion and attempts to achieve out-of-equilibrium gains through cooperation cause deviations from equilibrium.[8] It is worthwhile to point out that – in the presence of attempts at achieving cooperation – property (*v*) is likely to speed up equilibrium adjustment. Since cooperation attempts may compensate the decrease in adjustment speed that is due to money illusion, property (*v*) renders it more difficult to detect the impact of money illusion. However, if we observe that money illusion causes monetary non-neutrality despite the potential countervailing force of cooperation an even stronger result has been found.

Finally, property (*vi*) is likely to increase adjustment speed because it increases the chances that subjects find the equilibrium: The more methods are available for finding the equilibrium, the higher the chances that it will be found.

2.2. Experimental procedures and parameters

All major experimental parameters and design features are summarized in *table* 1. The experiment was conducted in a computerized laboratory with a group size of n = 4 subjects. In each group there were two types of subjects: Subjects of type x and subjects of type y. The payoff function differed among types. This difference implied that x-types had to choose a relatively low price in equilibrium while y-types had to choose a relatively high price (see *table* 1 for details). In the pre-shock phase of each treatment the money supply was given by $M_0 = 42$ while in the post-shock phase it was given by $M_1 = M_0 / 3 = 14$. In the pre-shock equilibrium the average price over all n group members is $\overline{P}_0^* = 18$ while in the post-shock equilibrium it is $\overline{P}_1^* = 6$.

8. In pilot experiments a price-setting game with an inefficient equilibrium (which is typical for monopolistic competition) has been implemented. However, it turned out that subjects quickly realize that there are out-of-equilibrium cooperative gains to be made. It is well known from many public good experiments (see e.g. LEDYARD 1995) that the adjustment process to Nash equilibria with free-riding is severely retarded, if not prevented, by subjects' attempts to achieve cooperative out-of-equilibrium gains. In the pilot experiments the pre- as well as the post-shock phase equilibrium adjustment was strongly retarded by cooperation attempts.

Except for the pre-shock phase of the real treatment (RT, which is explained below) subjects had to choose an integer $P_i \in \{1, 2, ..., 30\}$ in each decision period. In addition, they had to provide an expectation about $\bar{P}_{-i}$ which we denote by $\bar{P}^e_{-i}$. Finally, subjects indicated their confidence about their expectation $\bar{P}^e_{-i}$ by choosing an integer from 1 to 6 where 1 indicated that the subject is "not at all confident" while 6 indicated that he or she is "absolutely confident". At the end of each period each subject was informed about the actual realization of $\bar{P}_{-i}$ and the actual real payoff π_i on a so-called outcome screen (see *figure* Figure 39: in appendix A1). In addition, the outcome screen provided information about the subject's past choices of P_i, past realizations of $\bar{P}_{-i}$ and past real payoffs π_i.

Subjects received the payoff information in matrix form. Appendix A2 contains the payoff matrices for all treatment conditions. The payoff matrix shows the real or the nominal payoff, respectively, for each feasible integer combination of $(P_i, \bar{P}_{-i})$. Since subjects' choice sets contained 30 elements the payoff matrix had a 30 x 30 dimension. To inform subjects about the payoffs of the other type, each subject also received the payoff matrix of the other type. This information condition was common knowledge. The presentation of payoffs in the form of a matrix made it particularly easy to find the best reply for any given $\bar{P}_{-i}$: The subject just had to look for the highest real or nominal payoff in the column associated with $\bar{P}_{-i}$.

At the end of period 20 the nominal shock was implemented in the following way: Subjects were publicly informed that *x*- and *y*-types will receive new payoff tables. These tables were based on $M_1 = M_0 / 3$. Again each type received the payoff table for his own and the other type. Subjects were told that, except for payoff tables everything else remained unchanged. They were given enough time to study the new payoff tables and to choose P_i for period 21.[9] This procedure guarantees that in period 21 subjects face an exogenous and fully anticipated negative nominal shock. At the beginning of period 21 it was common knowledge that the experiment will last for further 20 periods.

9. In total subjects were given 10 minutes to study the new payoff tables and to make a decision in period 21. Yet, almost all subjects made their decision several minutes before time had elapsed.

Table 1: Parameters of the experiment

		Nominal treatment	Semi-real treatment	Real treatment
All periods	Representation of payoffs	nominal ($P_{-i}\pi_i$)	real (π_i)	real (π_i)
	Smallest nominal accounting unit	constant (variation in real terms)	constant (variation in real terms)	adjusted to keep real accounting unit constant
	Slope of reaction function	positive (NTC) or negative (NTS)	positive (SRTC)	positive (RTC) or negative (RTS)
	Group size	$n = 4$	$n = 4$	$n = 4$
	Information feedback in period t	P_{-i}, π_i	P_{-i}, π_i	P_{-i}, π_i
	Real equilibrium payoff (pre- and post-shock, for both types)	40	40	40
Pre-shock (t = 1 - 20)	Choice variable	$P_i \in \{1, 2, ..., 30\}$	$P_i \in \{1, 2, ..., 30\}$	$P_i \in \{3, 6, ..., 90\}$
	Money supply M_0	42	42	42
	Average equilibrium price $\bar{P}^*$ and average equilibrium expectation for the *whole* group	18	18	18
	Equilibrium price for type x	9	9	9
	Equilibrium expectation $\bar{P}_{-i}^e$ for type x	21	21	21
	Equilibrium price for type y	27	27	27
	Equilibrium expectation $\bar{P}_i^e$ for type y	15	15	15
Post-shock (t = 21 - 40)	Choice variable	$P_i \in \{1, 2, ..., 30\}$	$P_i \in \{1, 2, ..., 30\}$	$P_i \in \{1, 2, ..., 30\}$
	Money supply M_1	14	14	14
	Average equilibrium price $\bar{P}^*$ and average equilibrium expectation for the *whole* group	6	6	6
	Equilibrium price for type x	3	3	3
	Equilibrium expectation $\bar{P}_{-i}^e$ for type x	7	7	7
	Equilibrium price for type y	9	9	9
	Equilibrium expectation $\bar{P}_{-i}^e$ for type y	5	5	5

A total of 10 experimental sessions was conducted in fall and winter 1996/97 at the computerlab of the Institute for Empirical Research in Economics at the University of Zurich. The following table provides an overview.

Table 2: Dates, number of subjects, treatments

Date	Number of subjects	Representation	Slope of reaction function	Label
3.9. 1996	24	Nominal	Complements	NTC
4.9. 1996	20	Nominal	Complements	NTC
2.12. 1996	24	Nominal	Substitutes	NTS
6.12. 1996	24	Nominal	Substitutes	NTS
11.9. 1996	16	Semi-real	Complements	SRTC
13.11. 1996	24	Semi-real	Complements	SRTC
5.9. 1996	20	Real	Complements	RTC
12.9. 1996	20	Real	Complements	RTC
15.1. 1997	24	Real	Substitutes	RTS
16.1. 1997	20	Real	Substitutes	RTS

We recruited undergraduate students from all faculties (except the economics department) of the University of Zurich and the Federal Polytechnic School (ETH). The subjects were recruited partly in class, and partly by telephone from our subject data base. Subjects were sent a confirmation sheet indicating time and place of the experiment. In addition, we called subjects the evening before the experiment to remind them.

After welcoming the subjects at the Institute, we sat them at computer work stations which were separated by view protection boards to make sure that subjects made their decisions entirely privately. In addition, subjects were told that they would be excluded from the experiment and not entitled to any payments if they violated any of the rules of the experiment (e.g. no communication). Fortunately, we were never forced to take such disciplinary measures.

Subjects had to solve several exercises before the experiment and were all equipped with a pocket calculator. After having checked that all subjects answered the control questions correctly subjects were publicly informed that everyone solved the exercises correctly and read the summary page aloud to assure the subjects that everyone got the same instructions.

At the end of period 20, we implemented the change in the quantity of money by distributing new payoff tables. The pre-shock payoff tables were not collected but remained with the subjects throughout the post-shock phase. We announced publicly that everything else remained unchanged, specifically group composition. Subjects were given enough time (10 minutes in total) to decide on the prices they want to set in period 21. They could not make their decision within the first seven minutes. After this compulsory time span for studying the new tables had elapsed, most subjects immediately took their decisions.

Subjects were paid sFr. 15 (approx US$ 12) for showing up. Average total earnings were around sFr. 35 (approx. US$ 28). Earnings were paid out immediately after the experiment. Subjects had to confirm the amount received by signing a receipt.

2.3. Description of treatments

To test the hypotheses formulated in Part I, chapter 5 and to isolate factors causing money to be non-neutral, different treatments were implemented. Controlled variations of the environment are implemented along two dimensions: Section 2.3.1. discusses variations in the nominal vs. real representation or "framing", and section 2.3.2. describes the variations in the strategic properties of the decision environment.

2.3.1. Variation of representation

This section is an exposition of how the nominal vs. real representation of the environment was implemented. Three representations are discussed: A) the real, B) the semi-real, and C) the nominal representation.

A) Real representation

Figure 4 displays a part[10] of the pre- and post-shock payoff-matrices in the Real Treatment (RT). In the RT tables show real payoffs and the smallest nominal accounting unit is adjusted in such a way to hold the real accounting unit constant (see *table* 1). This adjustment of the smallest nominal accounting unit had been implemented to avoid money illusion induced by the FISHER effect. In the pre-shock phase of the RT subjects had to choose a price from the set $\{3, 6, \ldots, 90\}$. This means that the smal-

10. The whole payoff table is a 30 x 30 matrix. See appendix A2.

lest *nominal* accounting unit was 3 in this phase. In the post-shock phase, subjects had to choose any price from the set $\{1, 2, ..., 30\}$.[11] Hence, the smallest nominal accounting unit was 1 in the post-shock phase. This change in the smallest nominal accounting is necessary to keep the smallest *real* accounting unit constant between the pre- and the post-shock phase of the RT: At a money supply of M_0 a change in P_i by $\Delta P_i = 3$ has the same effect on the real payoff as a change of $\Delta P_i = 1$ when the money supply is $M_1 = M_0/3$.

Figure 4: Representation of the monetary shock in the Real Treatment (RT)

Pre-shock: Income table for firms of type *x*

Average price of other firms

selling price	3	6	9	
3	18	26	35	
6	12	18	25	
9	9	12	18	

Post-shock: Income table for firms of type *x*

Average price of other firms

selling price	1	2	3	
1	18	26	35	
2	12	18	25	
3	9	12	18	

Since the smallest nominal accounting unit has been adjusted equiproportionately to the change in the quantity of money, *the pre- and post-shock tables look exactly the same*, except for the headpieces of the table. For example, if subject i chooses a price of $P_{i0} = 3$ before the shock, and all other subjects in the group choose prices which yield an average of $P_{-i0} = 6$, subject i will get a payoff of $\pi_{i0}(3, 6, M_0) = 26$ (see upper diagram of *figure 4*). If all subjects adjust their nominal prices equiproportionately to the monetary shock, i.e. $P_{i1} = 1$ and $P_{-i1} = 2$, subject i (as well as all other subjects) will get the same real payoff as before the shock, i.e. $\pi_{i1}(1, 2, M_1) = 26$ (see lower diagram of *figure 4*). As a consequence, it should be completely transparent in

11. Note that in the post-shock phase the set of nominal choice variables is given by $\{1, 2, ..., 30\}$ *in each treatment* presented below.

this treatment (i.e. does not induce money illusion) to the subjects that the underlying economic structure has not been affected at all after the monetary shock. This treatment can be interpreted as a "currency reform". In a currency reform, one usually switches from 100 old currency units to 1 new currency unit. In this treatment, we switch from 3 old to 1 new currency unit.

B) Semi-real representation

The Semi-Real Treatment (SRT) differs from the RT insofar, as the smallest accounting unit is *not* adjusted equiproportionately to the monetary shock. Hence, subjects had to choose a nominal price from the set $\{1, 2, ..., 30\}$ before *and* after the monetary shock. *Figure 5* illustrates the effect of a change in the quantity of money when the smallest nominal accounting unit is held constant. The shaded cells in the pre-shock matrix (upper diagram) correspond to the cells shown in the post-shock matrix (lower diagram). If, for example, subject *i* chooses a price of $P_{i0} = 3$ and all other subjects choose on average prices of $P_{-i0} = 6$ before the shock, subject *i* will get a real payoff of $\pi_{i0}(3, 6, M_0) = 26$. If – after a reduction of the quantity of money of 3:1 – all subjects change their nominal prices equiproportionately, subject *i* (as well as all other subjects) will get the same payoff as before ($\pi_{i1}(1, 2, M_1) = 26$).

The representation effect explained above arises because of the fundamental function of money as unit of accounting and measurement of economic transactions. The FISHER effect is operative in the matrix as a kind of "zoom effect" because by keeping the smallest nominal accounting unit constant, we get a different representation of the same real economic conditions after a change in the quantity of money. Comparison of the behavior in the RT and the SRT allows to test for the FISHER hypothesis of money illusion.

Figure 5: Representation of the monetary shock in the Semi-Real Treatment (SRT)

Pre-shock: Income table for firms of type *x*

Average price of other firms

selling price	1	2	3	4	5	6	7	8	9
1	18	20	23	26	29	32	35	37	39
2	16	18	20	23	26	29	32	35	37
3	14	16	18	20	23	26	29	32	35
4	12	14	16	18	20	23	25	29	32
5	11	12	14	16	18	20	23	25	28
6	10	11	12	14	16	18	20	23	25
7	9	10	11	12	14	16	18	20	22
8	8	9	10	11	12	14	16	18	20
9	7	8	9	10	11	12	14	16	18

Post-shock: Income table for firms of type *x*

Average price of other firms

selling price	1	2	3
1	18	26	35
2	12	18	25
3	9	12	18

C) Nominal representation

The only difference between the Nominal Treatment (NT) and the SRT concerns the representation of payoffs: In the NT, each subject is given a payoff table displaying *nominal* payoffs. In the NT (as in the SRT), the smallest nominal accounting unit is held constant through *M*-levels. Subjects know that they are paid out according to their *real* payoffs at the end of the experiment. Of course, all subjects were instructed how to calculate real payoffs from nominal ones (see instructions in appendix A1). Subjects learned that real payoffs are obtained from nominal ones by dividing the latter by the actual average price, i.e by deflating nominal payoffs by the price level. Subjects had to solve several control questions where they had to correctly calculate real payoffs from nominal ones before the start of the experiment. It was common knowledge that all participants were able to answer the control questions correctly. In addition, we made sure that every single participant disposed of a pocket calculator.

We emphasize this procedure because it makes it highly implausible that subjects suffered from first-order money illusion in the NT. However, as the previous discussion has shown (see Part I, chapter 4), the absence of money illusion is not sufficient to rule out that money illusion matters. Instead, the absence of money illusion has to be common knowledge.

Figure 6: Representation of the monetary shock in the Nominal Treatment (NT)

Pre-shock: Income table for firms of type *x*

Average price of other firms

selling price	1	2	3	4	5	6	7	8	9
1	18	40	68	102	143	191	243	298	351
2	16	35	60	91	128	172	222	278	335
3	14	31	53	80	113	153	200	254	312
4	12	28	47	71	100	136	178	228	285
5	11	24	42	63	88	120	158	203	256
6	10	22	37	56	78	106	140	180	229
7	9	20	33	49	69	94	123	159	202
8	8	18	30	44	62	83	109	141	179
9	7	16	27	40	55	74	97	125	158

Post-shock: Income table for firms of type *x*

Average price of other firms

selling price	1	2	3
1	18	51	104
2	12	35	76
3	9	24	53

Figure 6 illustrates how the monetary shock is represented in the NT. In this treatment, the "veil of money" has two components which are jointly operative. Since we hold the smallest nominal accounting unit constant (integer prices from 1 to 30 can be chosen before and after the shock), the matrix looks "compressed". As in the SRT, the shaded cells in the upper diagram correspond to the payoffs shown in the lower diagram. But this fact is less obvious in this treatment than in the SRT, since subjects additionally have to deflate payoffs. To use the same example as above ($P_{i0} = 3, \overline{P}_{-i0} = 6$), subject i gets a nominal payoff of $\Pi_{i0}(3, 6, M_0) = 153$, which corresponds to a real payoff of $\pi_{i0}(3, 6, M_0) = \Pi_{i0}/\bar{P}_{-i0} = 153/6 = 26$ (rounded).[12] If all subjects adjust their prices equiproportionately, i.e. $P_{i1} = 1, \bar{P}_{-i1} = 2$, subject i (as well as all other subjects) gets the same payoff as before, i.e. $\pi_{i1}(1, 2, M_1) = \Pi_{i1}/\bar{P}_{-i1} = 51/2 = 26$.

It is worthwhile to emphasize that the NT and the SRT exclusively differ with respect to the payoff information. While in the NT the entries in the payoff table represent nominal payoffs $\bar{P}_{-i}\pi_i$, the entries in the SRT represent π_i. Thus to compute the real payoff for a particular $(P_i, \bar{P}_{-i})$-combination in the NT a subject just had to divide $\bar{P}_{-i}\pi_i$ by $\bar{P}_{-i}$. Note that nominal payoffs tend to be the higher the higher the price level $\bar{P}_{-i}$ (see *figure* 6). If subjects mistakenly interpret high nominal payoffs as high real payoffs, they are said to be prone to "high-number illusion".

Summary of variations in representation. *Table* 1 summarizes the most important differences among the treatments varying the representation of the economic environment. Of the two elements of inducing money illusion (i.e. the FISHER effect, and the deflationing effect), none is implemented RT, one is implemented in the SRT, and both are implemented in the NT.

Since in the RT the experimenter has, so to speak, removed the veil of money, it is more likely that subjects understand that nothing real has changed *and* that this is common knowledge. In contrast, in the other two treatment conditions subjects themselves have to pierce the veil of money, and have to form beliefs about whether other subjects are capable of doing so, and so forth. Therefore, the NT and the SRT capture elements of the "monetary veil in which most business transactions are shrouded".

Note that the tables are *identical* in the SRT and the RT after the shock (i.e. they are the same and look the same). In addition, the post-shock *payoff tables are exactly the same in real terms in all three treatments*.

12. Subjects were told in the instructions that average prices and real payoffs can only take integer values and that the usual rounding rules apply.

2.3.2. Variation of the strategic property

This subsection discusses the variation of the strategic properties of the economic environment. In the strategic complements condition, the reaction function has a positive slope, in the strategic substitutes case the reaction function has a negative slope. *Figure 7* shows how this controlled *ceteris paribus* variation works.[13] The *best reply function* for subject *i* can be found by taking average prices (columns) as given and looking for the prices P_i that yield the highest payoffs (shaded cells[14]). For example, if all other players choose prices such that the resulting average price is $\bar{P}_{-i} = 8$, subject *i*'s best reply is to choose a price of $P_i^{br}(\bar{P}_{-i}) = 11$ in the upper diagram. In this example, the reaction function has positive slope (the higher the average price of other subjects, the higher the payoff maximizing price for subject *i*), i.e. the environment is characterized by *strategic complementarity* in the upper diagram. The lower diagram depicts the case of *strategic substitutes*, where the slope of the reaction function is negative. *Note that the structure of payoffs is otherwise the same* in both treatments. For a given $\bar{P}_{-i}$ the payoff function is symmetric[15] and bell-shaped, and the payoff steps are the same. Thus, the only thing that is changed between these two treatments is the slope of the reaction function (see appendix A3).

The equilibrium value in *figure 7* is at a price of $P_i^* = 9$ and at an average price of $\bar{P}_{-i}^* = 5$. As can be seen, the reaction function is "flattened" around the equilibrium. This property has been implemented to make equilibrium play more stable. The stabilizing effect arises because the equilibrium choice is a best reply to several average prices. Hence, once subjects are in equilibrium, small deviations of the other *-i* players in the group from the equilibrium choice do not provide incentives for *i* to change the equilibrium choice.

---

13. The figure shows the post-shock tables in the SRT for *y*-type players. For other tables, see appendix A2.
14. Of course, experimental subjects were given payoff tables *without* shaded cells (see appendix A2).
15. The property that the profit function is symmetric in $\bar{P}_{-i}$ cannot be seen very easily. This is so because we used a continuous profit function which may have its maximum for a given $\bar{P}_{-i}$ at a non-integer P_i. Since we show only (rounded) profits for integer prices, it is not obvious that we in fact hold the shape of the profit function constant and just shift its maximum to higher prices for *i* at higher values of $\bar{P}_{-i}$.

Figure 7: Strategic complements vs. strategic substitutes (SRT)

Strategic complements

selling price \ Average price of other firms	1	2	3	4	5	6	7	8	9	10	11	12	13
1	9	7	5	4	4	4	3	3	2	2	2	1	1
2	12	9	7	6	5	5	4	3	3	2	2	2	1
3	18	13	9	7	7	6	5	4	3	3	2	2	2
4	26	18	13	10	9	8	6	5	4	3	3	2	2
5	35	26	19	14	12	11	9	7	5	4	3	3	2
6	40	36	27	20	18	16	12	9	7	5	4	3	3
7	35	40	36	29	26	23	17	12	9	7	5	4	3
8	25	34	40	38	35	32	25	17	12	9	7	5	4
9	18	25	34	39	40	39	34	25	18	12	9	7	5
10	12	17	25	32	35	38	40	34	25	18	12	9	7
11	9	12	17	23	26	29	36	40	35	26	18	12	9
12	7	9	12	16	18	20	27	36	40	35	26	18	12
13	5	7	9	11	12	14	19	26	35	40	35	25	18

Strategic substitutes

selling price \ Average price of other firms	1	2	3	4	5	6	7	8	9	10	11	12	13
1	2	3	3	4	4	4	5	7	9	12	17	25	34
2	3	3	4	5	5	6	7	9	12	18	25	34	39
3	3	4	5	6	7	7	9	13	18	25	34	39	35
4	4	5	6	8	9	10	13	18	26	35	39	35	25
5	5	7	9	11	12	14	19	26	35	40	35	25	18
6	7	9	12	16	18	20	27	36	40	35	26	18	12
7	9	12	17	23	26	29	36	40	35	26	18	12	9
8	12	17	25	32	35	38	40	34	25	18	12	9	7
9	18	25	34	39	40	39	34	25	18	12	9	7	5
10	25	34	40	38	35	32	25	17	12	9	7	5	4
11	35	40	36	29	26	23	17	12	9	7	5	4	3
12	40	36	27	20	18	16	12	9	7	5	4	3	3
13	35	26	19	14	12	11	9	7	5	4	3	3	2

Figure 8: Strategic complements vs. strategic substitutes (NT)

Strategic complements

selling price \ Average price of other firms	1	2	3	4	5	6	7	8	9	10	11	12	13
1	9	14	16	17	20	23	22	21	20	19	18	17	16
2	12	18	21	22	26	29	28	26	24	22	21	19	18
3	18	25	28	29	34	37	35	32	29	27	25	23	21
4	26	36	39	40	45	49	45	40	36	33	29	27	24
5	35	52	56	56	62	67	60	53	46	40	36	32	29
6	40	71	81	81	89	95	83	70	60	51	44	39	34
7	35	80	109	115	129	137	119	97	80	66	56	48	42
8	25	68	119	151	176	193	172	139	111	89	73	61	52
9	18	50	102	157	200	236	237	201	158	123	98	79	66
10	12	34	74	129	176	226	279	275	229	177	136	107	86
11	9	24	51	92	129	173	253	318	312	255	195	148	115
12	7	18	36	63	89	121	189	284	357	348	281	212	160
13	5	13	26	45	62	84	131	210	316	395	382	305	229

Strategic substitutes

selling price \ Average price of other firms	1	2	3	4	5	6	7	8	9	10	11	12	13
1	2	5	10	15	20	26	37	54	81	123	192	300	440
2	3	6	12	19	26	34	49	73	112	176	277	410	506
3	3	8	15	25	34	44	66	102	161	254	379	470	449
4	4	10	19	33	45	60	91	145	232	347	433	416	331
5	5	13	26	45	62	84	131	210	316	395	382	305	229
6	7	18	36	63	89	121	189	284	357	348	281	212	160
7	9	24	51	92	129	173	253	318	312	255	195	148	115
8	12	34	74	129	176	226	279	275	229	177	136	107	86
9	18	50	102	157	200	236	237	201	158	123	98	79	66
10	25	68	119	151	176	193	172	139	111	89	73	61	52
11	35	80	109	115	129	137	119	97	80	66	56	48	42
12	40	71	81	81	89	95	83	70	60	51	44	39	34
13	35	52	56	56	62	67	60	53	46	40	36	32	29

Figure 8 shows the payoff-tables in the nominal representation (for type *y*, post-shock). As in *figure* 7, the best replies are shaded. Of course, the best reply structure remains unaffected by the nominal representation. In addition, the cognitive complexity to find the best reply (and the equilibrium) is exactly the same in the two representations. A subject just has to look for the highest (nominal or real) payoff for any given $\bar{P}_{-i}$. As explained above, high nominal payoffs naturally tend to prevail at high price levels. Note, however, that high nominal payoffs tend to prevail at high individual prices P_i in the NTC (upper diagram of *figure* 8), but at relatively *low* prices P_i in the NTS (see lower diagram). This is the case because the highest real payoff for a given price level $\bar{P}_{-i}$ by definition prevails along the best reply and because the best replies have different slopes in the two treatments.

Strategic properties and the aggregate effects of first-order money illusion. The two aspects of representation inducing money illusion may affect the behavior of naive agents differently in the two treatments. In the presence of super-rational agents, the aggregate effects of money illusion depend on which aspect of money illusion dominates (see also Part I, chapter 4B). If a naive subject is prone to high-number illusion in the sense that he generally interprets high nominal payoffs for high real payoffs, it will tend to choose relatively high prices in the NTC, but relatively low prices in the NTS. Assume there is only one super-rational subject (who is free from high-number illusion) and *all other* subjects prone to first-order high-number illusion. In this case, the super-rational subject would tend to choose high (above equilibrium) prices in both treatments. As a consequence, aggregate nominal prices would be far above the equilibrium in the NTC, but be below the equilibrium in the NTS. What happens if all naive subjects are prone to the FISHER effect? In this case, the naive subjects would choose high (i.e. pre-shock equilibrium) prices. In reaction to this, the single super-rational would choose high prices in the NTC, but low prices in the NTS. As a consequence, aggregate nominal prices would be above equilibrium in both treatments, but prices would tend to be more sticky in the NTC than in the NTS. To summarize, both high-number illusion *and* nominal anchoring draw the choices of both naive *and* sophisticated agents to high prices, i.e. away from equilibrium in the NTC. In contrast, the behavior of naive and super-rational agents depends on the source of money illusion in the NTS. Given that H_{A3} holds, and given some mixture of the two motives and types, one would expect prices relatively far above the equilibrium in the NTC, but relatively close to the equilibrium in the NTS.

Negative vs. positive monetary shock and the effects of money illusion. The FISHER effect (or, nominal anchoring) effect would in general predict an inertial reaction, no matter what the direction of the shock, and no matter which strategic properties prevail. The behavioral impact of high-number illusion, on the other hand, depends on the direction of the shock.[16] Given strategic complementarity and a positive monetary shock, people prone to high-number illusion would tend to increase their nominal prices a lot (because high nominal payoffs prevail at high prices). With a monetary contraction, they would tend to reduce them only a little. Since we do *a priori* not know which aspect of money illusion (if any) is more important, the implementation of a negative monetary shock allows to detect whether money illusion is a cause of monetary non-neutrality.

2.4. Advantages of the present experimental design in the investigation of monetary non-neutrality

As discussed in Part I, the empirical evidence from macroeconomic field data on the neutrality of money is controversial. This is the case because empirical research with field data suffers from important limitations: It is difficult to correctly measure aggregate variables, and this mismeasurement may matter for the conclusions drawn on the issue of monetary neutrality. In addition, truly causal relations are hard to establish because of the exogeneity problem. On the other hand, there is evidence from questionnaire studies, suggesting that money illusion is widespread at the individual level. What can we hope to learn from the present experimental study about the debate on the neutrality of money that we cannot learn from analyzing macroeconomic "happenstance" data or from questionnaire studies? The following paragraphs provide some answers.

Advantages over investigations with macroeconomic field data. An obvious advantage of experimentally generated data over macroeconomic field data consists in *correct measurement* of endogenous variables like aggregate prices ($\bar{P}$) and real economic activity (π). An important advantage of the experimental method is control over the environment and the informational conditions: We know the exact extent of a truly exogenous monetary shock and the *theoretical equilibrium values* (π^*, P^*, see *table* 1). This is allows to discern whether observed behavior is an equilibrium or out-of-equilibrium realization. Hence, one can judge whether behavior eventually converges to the equilibrium and, if so, how fast this convergence is. This feature is crucial in an investigation of nominal rigidity which by definition is a disequilibrium phenomenon. Exogenous variables, such as the quantity of money (*M*), are truly *exogenous* since

16. It also depends on strategic properties, as has been explained in the preceding section.

they are implemented by the experimenter. This feature allows to investigate whether or not the monetary shock causes real effects. Furthermore, we control what economic subjects know about the economic environment. This puts us in the position to implement an *anticipated* (or, for that matter, an unanticipated) monetary shock, where we can be sure that every subject knows how she and everybody else is affected by this change in the quantity of money. Subjects are not only asked to choose a (nominal) price in every period, but also had to indicate what average price they expect for the current period ($\bar{P}_{-i}^{e}$), and how confident they are that this expectation is correct. Consequently, this design allows to test for various rationality assumptions: Do subjects correctly predict the price level (i.e. do they have rational expectations)? How does money illusion affect these expectations? Given the subjects' expectations of the price level, do subjects behave in a "near-rational" manner (AKERLOF and YELLEN, 1985b, 1985b) or do they choose prices which are exact best responses? Does a nominal shock create uncertainty and thus undermine trust in the rationality of other subjects? As will be explained in Part III, these expectations data allow to investigate whether losses occur because subjects fail to correctly predict the behavior of others, or because they are "irrational" in the sense that they do not choose prices which are best replies given their expectation on aggregate behavior.

Experimental methods also allow us to be "unrealistic" in the sense that we can parallel theory with our decision environment. This enables us to actually *test the behavioral predictions of a theory.* The present study, thus, interprets macroeconomics as an indirectly experimental science (see chapter 1.C).

Isolating causes and consequences of money illusion. Money illusion means that behavior depends on whether the same objective situation is framed in nominal or in real terms. A particularly transparent example of money illusion is the case where people behave differently when they receive payoff information in real or in nominal terms. In business life, almost all business transactions involve nominal payoff information. To detect this kind of money illusion it would, thus, be necessary to find situations in which a real frame prevails. By comparing people's behavior under the nominal and the real frame one could isolate the behavioral impact of money illusion. Unfortunately, business life does not seem to provide examples in which the same objective situation is sometimes represented in nominal terms and sometimes in real terms. This is one important reason why we rely in our empirical examination on experimental methods. In the present context a major advantage of experimental methods is that the "frame" is under the control of the experimenter. Finally, experiments allow to *establish causal relations* through controlled *ceteris paribus* variations of the environment. By changing only one aspect of the environment and by comparing behavior in the respective treatments we are in the position to argue that this change in behavior must have come from this single variation. For example, we may change the (nominal) representation without changing the real economic incentive structure (or anything else). If behavior is significantly different under the two representations, these differences must have been caused by the representation.

Advantages over questionnaire studies. The questionnaire study discussed in Part I, section 3.1 by SHAFIR *et al.* (1997) shows that many people seem to be prone to money illusion. Why do we need an experimental study then? What are the advantages of experimental methods over questionnaire studies in investigating money illusion? First, economists tend to question whether the answers given in a hypothetical situation extend to an economically relevant situation. Respondents do not have an incentive to think hard about the answers since answering one way or the other does not have any material consequences for them. In contrast, economic experiments expose subjects to economic incentives. The income they earn in the experiment depends on the decisions they take. One may thus argue that what we observe in questionnaire studies are what people say they would do in some economic situation. What we observe in an economic experiment is what people actually do in an economically relevant situation. As everybody knows, what people say they would do and what they actually do is not always the same.

Second, economists tend to be interested in explaining *aggregate* behavior of some sort. They are not primarily interested in whether or not people are prone to money illusion, but whether this tendency has any repercussions on (interactive) economic decisions. One could argue that even if questionnaire studies have shown that people sometimes make mistakes in judging situations, this may still not matter in the aggregate since (unsystematic) effects may wash out with interaction. To the contrary, in economic experiments, we can study interaction and investigate whether individual biases translate to the aggregate level. The present study is not primarily designed to test whether or not people are prone to money illusion, but whether money illusion matters on the aggregate level under varying circumstances.

Part III

Results of Experimental Study

Chapter 1 tests the proposition that an anticipated monetary shock is generally neutral in the absence of exogenous frictions. The test is conducted in a "natural" environment in which strategic complementarity and a nominal representation prevails. The results show that the Neutrality hypothesis H_0 is clearly rejected. Besides establishing the crucial fact to be explained, chapter 1 also serves the purpose of explaining the presentation of data which is chosen throughout Part III. In addition, the analysis of chapter 1 suggests non-rational expectations as a tentative explanation of the causes of monetary non-neutrality. This tentative explanation serves to operationalize further hypotheses on expectation formation. These hypotheses are to be tested in chapters 2 and 3.

Chapter 2 investigates whether money illusion is a cause of monetary non-neutrality. First, we discuss the effects of variations in representation given strategic complementarity. In this case, the Money illusion hypothesis H_{A1} is supported and money illusion is shown to be a causal factor of monetary non-neutrality. When the environment is characterized by strategic substitutes, hypothesis H_{A1} appears to be incomplete, whereas the hypothesis H_{A3} receives support by the data.

Chapter 3 investigates whether strategic complementarity is a cause of monetary non-neutrality. That is, the effects of a variation of strategic properties of the decision environment are discussed. The HALTIWANGER / WALDMAN-hypothesis H_{A2} is supported when the environment is represented in nominal terms. When represented in real terms, hypothesis H_{A2} appears to be incomplete. It is shown that strategic properties matter more when money illusion can be supposed to be prevalent. Hence, the "Money illusion *cum* strategic anticipation" hypothesis H_{A3} is accepted.

Chapter 4 summarizes the results.

Chapter 1: Non-neutrality with strategic complementarity

"The main finding that emerged from the research of the 1970s is that anticipated changes and unanticipated changes in money growth have very different effects. Anticipated monetary expansions ... are ***not*** *associated with the kind of stimulus to employment and production that Hume described. Unanticipated monetary expansions, on the other hand, can stimulate production as, symmetrically, unanticipated contractions can induce depressions."*

ROBERT E. LUCAS (1996: 679)

According to the Neutrality hypothesis H_0 anticipated monetary shocks do not affect real economic activity in the absence of exogenous of informational frictions. This should hold irrespective of framing and irrespective of the strategic property of the environment. According to this hypothesis anticipated money should be neutral in all cells of *table* 3. This chapter shows that a fully anticipated monetary shock is massively short-run non-neutral when the environment is represented in nominal terms and strategic complementarity prevails (see shaded cell NTC in *table* 3). Thus, the Neutrality hypothesis H_0 is rejected. Yet, the proposition of long-run neutrality is supported by the data. Individual level data on price expectations, best reply behavior and a profit decomposition show that short-run non-neutrality mainly arises because of non-rational expectations and strategic interaction, whereas individually irrational behavior is of relatively minor importance.

Table 3: Nominal representation when the environment is characterized by strategic complementarity (NTC)

	Nominal representation	Semi-real representation	Real representation
Strategic complements	NTC	SRTC	RTC
Strategic Substitutes	NTS		RTS

The data presented in this chapter has been generated by the behavior of 44 subjects who earned approx. 27$ on average within roughly two hours.

The presentation of results proceeds as follows: Section A) shows that the monetary shock causes pronounced nominal rigidity and section B) shows that this translates into massive real income losses. As explained in Part II, all subjects simultaneously had to decide on three variables in each period: The nominal price they want to set, the average nominal price they expect in the current period, and how confident they are that their price expectation will prove to be correct. Section C) shows that subjects behaved in an individually rational manner given their price expectations. Section D) argues that expectations data are reliable, and discusses to what extent expectations were correct and in equilibrium. Section E) presents a loss decomposition, showing that individual real income losses resulted from incorrect expectations. Section F) shows that subjects' confidence in their ability to predict average prices was shaken by the monetary shock. Section G) summarizes chapter 1.

A) Nominal rigidity

Figure 9 shows the evolution of average nominal prices over all groups. As can be seen, nominal prices are very close to the theoretical prediction ($\bar{P}_0^* = 18$, indicated by the straight line) throughout the first half of the experiment (periods 1 to 20).[1] However, the main focus of the present study is how subjects react to a monetary shock when they already are in equilibrium. The pre-shock phase hence mainly serves the purpose of equilibrating the system and will in the sequel not be explicitly analyzed (for a survey on how subjects learn to find and play the equilibrium see e.g. MAILATH 1998). Since average behavior in fact equilibrated in all treatments before the shock, the presentation of results will henceforth only include the last three pre-shock periods as a point of reference.

1. The observation of quick convergence from the very beginning of the experiment remains somewhat puzzling. However, one may speculate that quick convergence in the pre-shock phase is due to high-number illusion, since the pre-shock equilibrium is at prices with very high nominal payoffs. Another explanation may be the following: Due to the FISHER effect (which results from holding the discrete smallest nominal accounting unit constant across M-levels) the equilibrium choice is a best reply to five average prices in the pre-shock phase but only to three average prices in the post-shock phase. This makes the equilibrium choice a less risky strategy and may induce more subjects to actually choose it in the pre-shock phase.

Figure 9: Evolution of average nominal prices (NTC)

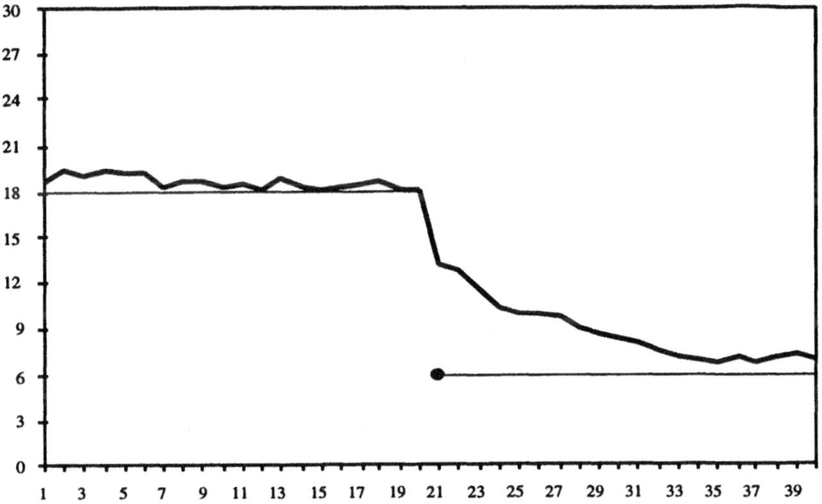

At the end of period 20, new decision tables were distributed and subjects were given enough time (10 minutes in total) to decide what prices to choose in period 21. Thus, the monetary shock was fully *anticipated* in the sense that subjects were perfectly informed about the shock. There were no exogenous frictions whatsoever, i.e. prices could be changed in any period at no cost and there were no informational problems. Therefore, in the presence of common knowledge of rationality subjects should have instantaneously and equiproportionately have adjusted nominal prices to the shock. According to H_0, we should observe $\bar{P}_1^* = 6$ in period 21 (see dot in *figure* 9). Even though subjects react to the monetary shock, aggregate nominal prices only slowly adjust to the new equilibrium level. Aggregate behavior exhibits massive nominal inertia during 12 periods after the shock.[2] Eventually, average nominal prices converge from above to the equilibrium value.

R1: *An anticipated monetary shock leads to massive nominal inertia in an environment which is represented in nominal terms and is characterized by strategic complementarity.*

2. The hypothesis that average group prices are in equilibrium cannot be rejected for the three periods before the shock. For the consecutive *12 periods* the hypothesis can be rejected at the 5% level, using a one-tailed Mann-Whitney test. In periods 33-40 we are unable to reject the hypothesis that groups are in equilibrium.

B) Monetary non-neutrality

Does an anticipated monetary shock cause real effects? Does the nominal rigidity documented in result R1 translate into real income losses? The main finding with respect to monetary non-neutrality is summarized in

R2: *An anticipated monetary shock causes massive real effects in an environment which is represented in nominal terms and is characterized by strategic complementarity.*

To measure the real effects of the monetary shock, it is calculated by how much actual real income of group *j*, π_j, falls short of real income in equilibrium π^*. For this purpose $L_{jt} = (\pi_{jt} - \pi^*)/\pi^*$ is computed for each group *j* in each period *t*. L_{jt} is a measure of the income loss relative to the equilibrium payoff as a percentage of the equilibrium payoff. Since the equilibrium is efficient[3] it is also a measure of the *efficiency loss*. Average efficiency losses were close to zero in the three periods before the monetary shock, but amounted to 65% in period 21, 47% in period 22, and were still 35% in period 23 (see also *table* 2 in appendix A3 for details).[4] Average efficiency losses after period 35 increase again because of high variance in behavior of very few subjects, whereas most other subjects are in equilibrium. Therefore, efficiency losses of the *median group*, which is less sensitive to outliers, are shown in *figure* 10.[5] The evolution of the average efficiency loss of the median group has the same qualitative features as the simple average. However, it is more transparent with this series that efficiency losses towards the end of the experiment are again close to pre-shock levels. The efficiency loss of the median group remains stable at pre-shock levels from period 34 on.

One of the core propositions of modern macroeconomics is that in the long run money is neutral. In the absence of multiple equilibria and hysteresis effects, as described, for example in BLANCHARD and SUMMERS (1986), there are good theoretical reasons for long-run neutrality. Yet, we are not aware of *direct* and unambiguous evidence in favor of long-run neutrality.[6] In the context of our simple laboratory economy we get, however, clear evidence for long-run neutrality.

---

3. See Part II, section 2.1.
4. Statistical tests suggest that aggregation over types is unproblematic. *x*-type and *y*-type players do not have significantly different efficiency losses.
5. That is, group averages L_{jt} are ranked and the median value is shown. This measure will be used throughout Part III to facilitate comparison between treatments. Note that the identity of the median group may change from period to period.

Figure 10: Real efficiency loss of median group (NTC)

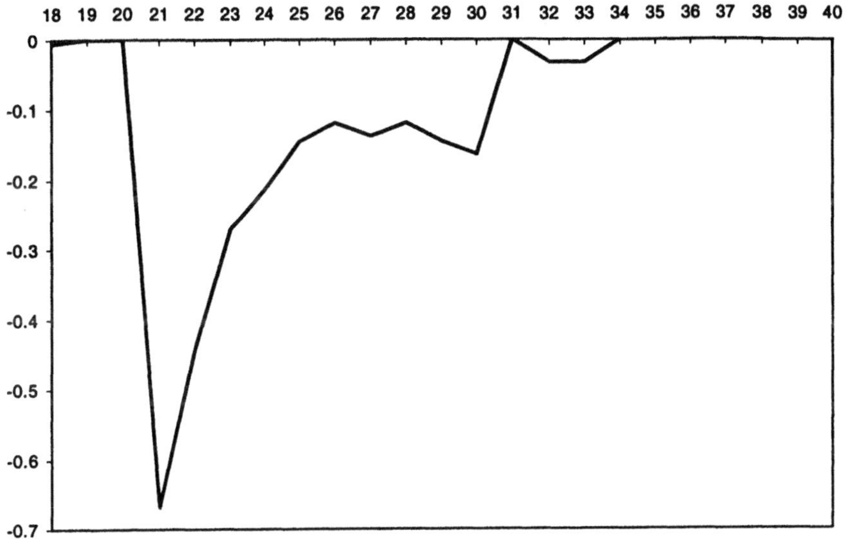

R3: *Despite massive short-run non-neutrality, the anticipated monetary shock is long-run neutral.*

In fact, we seem to have been able to answer KRUGMAN's question (see quotation in Part II, chapter 2): "how could you experimentally produce recessions and recoveries?" *Figure* 10 does not only show a (admittedly very stylized) recession, but it is also a recession that has been *caused* by an anticipated monetary contraction, which according to the New Classicals should not affect real economic variables.

Reasons for nominal inertia. In principle, there are several potential explanations for the substantial amount of nominal inertia as documented above.[7] First, as all experimentalists are aware, behavior of subjects usually does not perfectly and instanta-

6. BLANCHARD (1990: 828): "All the models we have seen impose long-run neutrality of money as a maintained assumption. This is very much a matter of faith, based on theoretical considerations rather than on empirical evidence." However, several empirical studies are on this issue have recently become available. FISHER and SEATER (1993) reject the long-run neutrality of money for the U.S. Yet, BOSCHEN and OTROK (1994) question these results.They argue that if one accounts for the exceptional period from 1930 to 1939 (when an extraordinary number of bank failures occurred) by a dummy variable, long-run neutrality prevails for U.S. data. HAUG and LUCAS (1997) provide independent evidence that the rejection of long-run neutrality by FISHER and SEATER is based on the anomalous period of the 1930's. LUCAS (1996) provides data suggesting that long-run neutrality is the rule rather than the exception. In contrast, BALL'S (1998) empirical study indicates that money shocks have long-run effects by changing the equilibrium unemployment rate.
7. Keep in mind that there are no exogenous nominal of informational frictions whatsoever present in this design.

neously adjust to some shock because there is some inevitable noise in behavior. This suggests that there may be some basic inertia which does not have anything to do with money illusion. This important aspect will be addressed in the chapters to follow by comparison of behavior across treatments. Second, subjects may simply be confused by the monetary shock and do no longer play best replies to their expectations. The near-rationality approach, for example, assumes that a fraction of subjects fails to maximize after a monetary shock because the losses associated with non-adjustment are small. In our experiment some subjects may also be somehow anchored at the "historically" given pre-shock price level so that they do not play best reply. Third, it may be that nominal inertia is caused by sticky expectations. Subjects may believe that the prices of other subjects remain relatively high and play best reply to this expectation. Due to strategic complementarity their own price will then also be relatively high. Section C) to E) provide a tentative explanation for the reasons of monetary non-neutrality.

C) Best reply behavior

This section analyzes whether subjects chose prices which were best replies *given* their expectations. For each subject *i* and each period *t* it is calculated whether P_{it} is the choice that maximizes π_{it} given the price expectation $\overline{P}^e_{-it}$. In general, subjects do choose prices which are exact best replies to the price expectation they indicate. As can be seen from *figure* 11, more than 70% of subjects do play best reply in all periods. However, the monetary shock seems to have influenced best reply behavior: The percentage of best replies drops from 95% to slightly below 70%, but quickly recovers to pre-shock levels. Since best reply behavior is (given some subjective price expectation) independent of strategic interaction, we interpret best reply behavior as an indicator of *individual rationality.*

> **R4:** *A large majority of subjects chooses exact best replies to $\overline{P}^e_{-it}$ in all periods. The percentage of best reply behavior slightly declines as a consequence of the monetary shock.*

Reasons for relatively high levels of individual rationality. Compared to what we know from other experimental studies on individual rationality, percentages of up to 95% perfectly rational behavior is high. There may be two reasons for this. First, the design makes the calculation of best replies particularly easy because subjects' payoff functions depend on the average price of *other* players $\bar{P}_{-i}$ which does not depend on the price choices of subject *i*. If the payoff function were made dependent on $\bar{P}$, i.e. including *i*'s price choice, finding a best reply would have been cognitively much more difficult. Second, subjects incurred considerable losses (approx. 10%) if they made the smallest possible best reply error of $\Delta P_i = \pm 1$. Subjects thus have a strong incentive to avoid such losses which certainly are not of second order. One may speculate that if conditions were implemented which favor "near-rational" behavior (i.e. deviations from optimality lead to *small* individual losses, see AKERLOF and YELLEN, 1985a, 1985b), this kind of behavior would have considerable repercussions on the aggregate.

Figure 11: Percentage of subjects choosing exact best replies to $\bar{P}^e_{-it}$ (NTC)

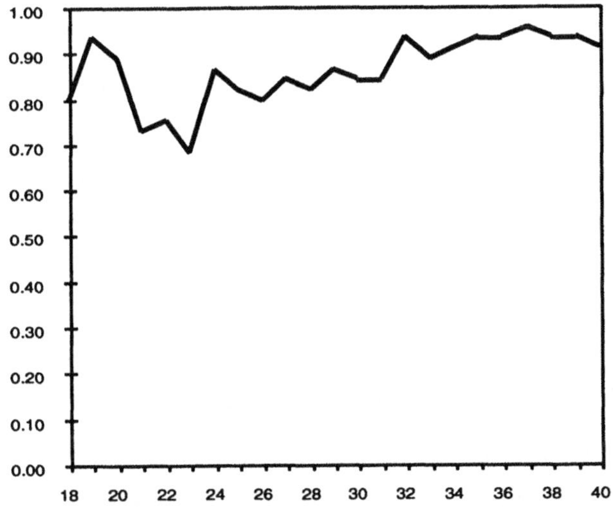

D) Price expectations $\bar{P}^e_{-i}$

Section C) has shown that subjects in general behave very rationally *given* their expectations. This section takes a closer look at expectations data. Three questions are discussed. First: Are expectations data reliable? Second: Are individual expectations objectively correct? Third: Do subjects expect other subjects to choose equilibrium prices, and how does the monetary shock affect this expectation?

Reliability. In each period *t* each subject *i* had to indicate which average price $\bar{P}^e_{-i}$ he or she expects. One may question the reliability of such an indication since subjects were not paid for indicating correct expectations. Hence, one may doubt that the reported expectations are a reliable measure of subjects' actual. expectations. There are good reasons for believing that the data on $\bar{P}_{-i}^{e}$ in fact is a fairly reliable measure of actual expectations because forming expectations that are as correct as possible are essential to decide which price to set. Thus, subjects in fact had an incentive to think hard about which average prices to expect. There are two facts suggesting that subjects reported expectations truthfully. First, as shown in section C), subjects in general chose prices which were best replies to the expectation they report. A rational subject would always want to choose a best reply to the true expectation but not to some other average price. The very high incidence of best reply behavior (see *figure* 11) suggests that expectations data are of high quality and thus reliable. A second fact suggesting that expectations data are reliable results from the analysis of correctness of expectations.

Correctness of expectations. Price expectations were objectively quite precise before the shock and from period 36 on, as can be seen in *figure* 12. If expectations data were not reliable, we would not observe up to 90 percent of subjects having correct expectations. In period 21, the first post-shock period, behavior seems to be much more difficult to predict (remember that the group composition remains constant, which was common information). The percentage of correct predictions drops from almost 80% to 5%, and only slowly returns to pre-shock levels. Only after 15 periods in the post-shock phase the percentage of correct expectations exceeds the pre-shock level.

Stickiness of expectations. The average expectation over all subjects for the first post-shock period was $\bar{P}_{-i}^{e} = 14$ which is far above the equilibrium expectation of $\bar{P}_{-i}^{*} = 6$ (for a graph, see *figure* 21). Thus, price expectations were not only incorrect, but also very sticky. This means that subjects expected *other* subjects (in period 21) to adjust their prices slowly. Note that subjects expected other subjects to choose relatively high prices before they knew how others are going to react to the monetary shock.

R5: *Reported expectations on average prices seem to be reliable. After the shock, aggregate behavior is much more difficult to predict and a large majority of subjects holds sticky (disequilibrium) expectations.*

Figure 12: Percentage of subjects with correct expectations $\bar{P}_{-i}^{e}$ (NTC)

(Dis-)equilibrium price choices. *Figure* 13 illustrates that the anticipated monetary shock is a disequilibrating force. In the last pre-shock period almost all (89%) subjects chose equilibrium prices, in the first post-shock period this only applies to a small minority (11%). This dramatic change in individual behavior also translates to the aggregate where average nominal prices exhibit massive nominal inertia (see *figure* 9). The percentage of subjects choosing equilibrium prices qualitatively tracks the evolution of efficiency losses (compare *figures* 11 and 13). This finding is not unexpected since the design implements an efficient equilibrium. The relation between (disequilibrium) price choice and real income losses will be discussed in detail in section E).

Figure 13 allows to discuss the effects of strategic complementarity (see chapter 3 of this Part for an extended discussion). The dramatic decline of equilibrium price choices from 89% to 11% in the wake of the monetary shock can be explained by either assuming that many subjects did simply not understand how to adjust their nominal prices ("first-order money illusion") or by assuming that relatively few subjects suffered from first-order illusion, but the many rational subjects anticipated that there will be some first-order illusion, and thus chose non-(above-)equilibrium prices. It seems that some subjects overestimated the degree of rationality of other players in period 21, since the percentage of equilibrium prices drops even further from 11% (period 21) to 5% in period 23. This result can be interpreted as follows: At least 11%

of subjects perfectly understood how they should adjust their nominal prices to the money shock, given that everybody else would perfectly adjust. Unfortunately, these subjects had to suffer large losses in the first few post-shock periods because they held incorrect price expectations (they did not expect the other subjects to choose disequilibrium prices). To illustrate, in period 21, subjects who chose the equilibrium price and had to bear a real income loss of 80%, whereas the subjects who did not choose equilibrium prices incurred a smaller income loss of 63% on average. In view of these losses, most subjects who chose equilibrium prices in period 21 decided to increase their nominal prices in period 22. Hence, the percentage of subjects choosing equilibrium prices dropped further after period 21.

Figure 13: Percentage of subjects choosing equilibrium prices (NTC)

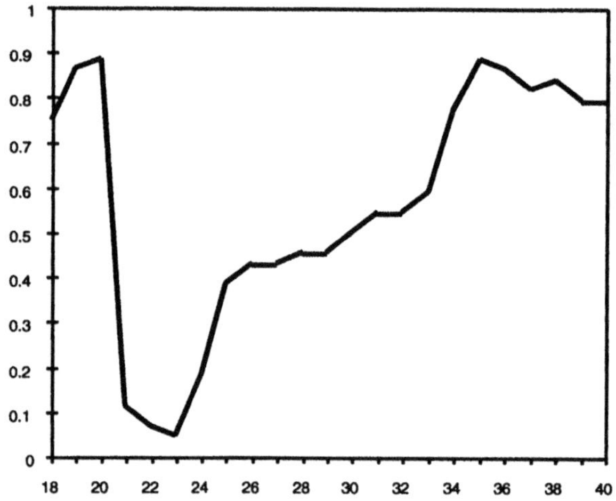

A tentative explanation. The analysis of individual level expectations allows to provide a first tentative explanation of massive nominal inertia and monetary non-neutrality. The main reason seems to be that the monetary shock systematically affects expectation formation. We have seen above that impact expectations were extremely sticky after the monetary shock. Section C) has shown that subjects predominantly choose best replies to their expectations. Given disequilibrium expectations, best reply behavior leads to disequilibrium price choices. In particular, given strategic complementarity, sticky (above equilibrium) expectations and best reply behavior lead to sticky aggregate prices. Since the implemented equilibrium is efficient, sticky (disequilibrium) price choices lead to efficiency losses. The next section provides some more evidence on this chain of argumentation. However, the discussion in this chapter leaves open the deep question of why expectations were sticky. Chapter 2 of this Part provides some answers.

> ***R6:*** *Sticky expectations (see R5) combined with widespread best reply behavior (see R4) lead to sticky (disequilibrium) price choices and sticky aggregate price adjustment. Sticky aggregate nominal prices after a negative monetary shock cause efficiency losses.*

E) Loss decomposition

With the additional information on expectations and best reply behavior discussed above, the task of loss decomposition can be addressed. This procedure provides additional insights as to why real income losses occurred. Nominal rigidity and monetary non-neutrality after an anticipated shock obviously are pervasive in this environment as stated in results R1 and R2. Chapters 2 and 3 of this Part isolate causal factors of monetary non-neutrality by controlled *ceteris paribus* variation of the decision environment. This section investigates why efficiency losses occurred at the individual level by decomposing total losses into three categories. We proceed by first describing the procedure, and then presenting results.[8]

Description of loss decomposition procedure. Losses may arises for three reasons in the present design. First, a subject may incur losses although she chooses a best reply on her expectation if she fails to correctly predict other subjects' behavior. Such "losses from failure to correctly predict average prices" (L^{Exp}) may arise because the subject does not have the cognitive ability to predict the (in principle easily predictable) behavior of others or because other subjects in the group act in an unpredictable manner. Second, a subject may correctly anticipate the average price but fail to act rationally on this (correct) expectation.[9] This source of real losses will be labeled "losses from failure to play best reply to subjective price expectations" (L^{BRF}). This kind of loss depends exclusively on individual rationality of behavior, and does not depend on strategic interaction with other subjects in the group. Finally, a subject may incur losses even though she correctly predicts other subjects' behavior and acts perfectly

8. For a more detailed discussion of the procedure and complete results, see appendix A3.
9. Subjects could only report one value for their price expectation $\bar{P}^e_{-i}$. If subjects are uncertain about the behavior of others, their expectations may be described by some distribution over $\bar{P}^e_{-i}$. Choosing a best reply to the reported value of $\bar{P}^e_{-i}$ (which can be thought of as a point estimator) is rational for all symmetric distributions over $\bar{P}^e_{-i}$.

rational. These losses will be labeled "losses from aggregate demand externality" (L^{ADE}). These three kinds of losses are illustrated in *figure* .Illustration of loss decomposition

Figure 14: Illustration of loss decomposition

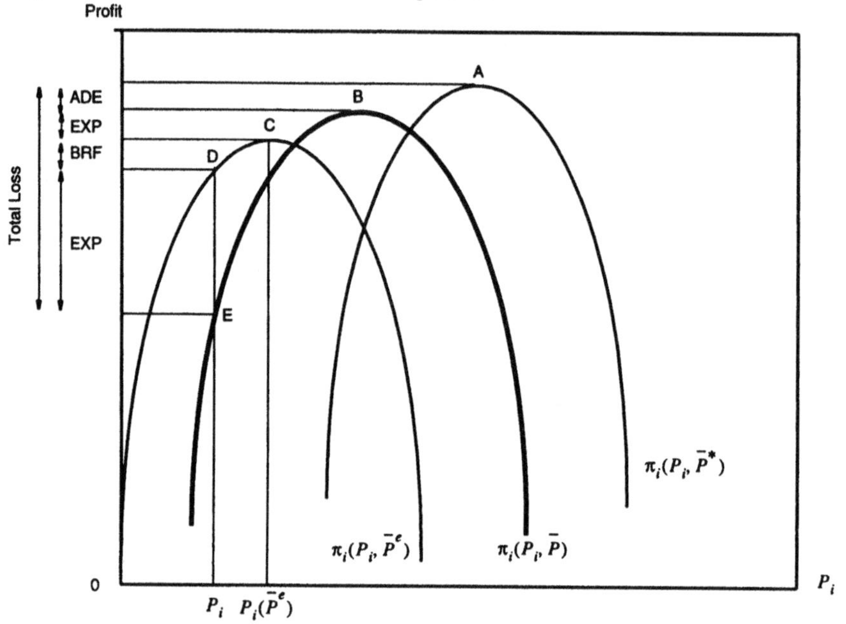

Suppose that subject i expects an average price of $\bar{P}^e$. This expectation gives rise to the (expected) profit function $\pi_i(P_i, \bar{P}^e)$ (see left parabola in *figure* 14). If this subject wants to maximize her payoff (given her expectation on the average price), she will choose the price $P_i(\bar{P}^e)$, which is the best reply given her expectation of $\bar{P}^e$. Suppose that this subject i is not fully rational for some reason and chooses a price P_i which is not a best reply on her expectation. If her expectation of $\bar{P}^e$ turns out to be correct, this subject will receive a lower payoff because of her failure to play best reply on her expectation (compare points C and D). Now suppose that subject i's expectation is not correct, and an average price of $\bar{P}$ materializes. This gives rise to the bold faced payoff function $\pi_i(P_i, \bar{P})$. If subject i had the correct expectation and was playing best reply on this expectation, she would have received a payoff indicated by the point B. But since she chose the price P_i, she will in fact get a low payoff indicated by point E in the diagram. The loss the subject incurs (B - E) can be partially attributed to the failure to play best reply on her expectation (C - D), partially to her failure to correctly predict the average price [(B - E) - (C - D)]. Finally, suppose that the actual average price $\bar{P}$ is not equal to the equilibrium price level $\bar{P}^*$. In equilibrium, subject *i* would have received a payoff illustrated by point A. The best a fully rational and perfectly foreseeing subject can do when the actual average price is $\bar{P}$ is to reach payoff B, which is smaller than the equilibrium payoff (point A). The difference A - B does not arise because the subject behaved irrationally or predicted incorrectly, but because *other* subjects chose prices that were not in equilibrium. Such losses thus constitute an aggregate demand externality.

Results from loss decomposition. Efficiency losses are small before the shock but increase dramatically after the shock (see *figure* 10 and *table* 7). In the first post-shock period total average efficiency loss is 65%. The loss decomposition shows[10] that 54% of efficiency are lost because of failure to correctly predict average prices, 9% are lost because some subjects did not play best reply, and 2% are lost due to aggregate demand externalities. Put differently, more than *83% of observed efficiency losses arise because subjects hold incorrect price expectations* in period 21. Thus, the lion's share of impact efficiency losses can be attributed to subjects' difficulties to correctly predict average prices. For example, In the first five post-shock periods (period 21-26) the median subject lost 87% of her profits because she had wrong expectations. Most subjects (80%) lost more than half of their profits in these periods because they had wrong expectations. Losses from aggregate demand externality and from failure to play best reply are generally much less important. For example, the average loss from best reply failure over all post-shock periods is less than 5%, the loss of the median subject due to best reply failure is below 1%. That is, some subjects incurred relative large losses (L^{BRF}), whereas most subjects almost perfectly played best reply. For example, the subject with the worst best reply behavior lost 30% of profits, whereas one third of all subjects (32%) did not have to bear any losses at all in the entire post-shock phase (periods 21-40).

R7: *The loss decomposition shows that efficiency losses after the anticipated monetary shock primarily occur because subjects hold incorrect expectations.*

---

10. For more detailed results see *table* 17 in appendix A3.

F) Subjective confidence in price expectation

Subjects also had to indicate how confident they were that their price expectation would be correct by choosing an integer from 1 to 6 (1 = "I am not at all confident that my expectation will be correct"; 6 = "I am very confident that my expectation on the average price will be correct", see instructions in appendix A1 for details). This expression of confidence can be interpreted as a *subjective* measure of correctness of expectations. As can be seen from *figure* 15, confidence (the subjective measure) qualitatively tracks the percentage of actually correct predictions (the objective measure, see *figure* 13). Note the breakdown of average confidence caused by the *announcement* of the monetary shock from almost 5.5 to about 3. (Subjects had to indicate their confidence in period 21 before knowing the consequences of the shock.)

One may wonder whether the subjective indication of confidence in one's price expectation is a good measure of expectational error. To check for this, Spearman correlations of absolute expectational errors and reported certainty have been calculated for each subject ($n = 44$) and every period (t = 18-40). The null hypothesis is that the two are unrelated, the alternative hypothesis is that they are negatively correlated. This null hypothesis can be rejected before the shock, but cannot be rejected for the three periods after the shock (and in period 29) at the 5% level of significance. We conclude from this that the subjective indication of confidence is in general a reasonably good measure of objective expectational error, but this does not seem to be true exactly for the periods of prime interest. After the anticipated shock subjects make objectively more mistakes in predicting aggregate behavior (see *figure* 12). In addition, subjects underestimate the worsening of the precision of their predictions. Thus, the monetary shock seems to have fundamentally shaken the process of expectation formation.

R8: *Confidence in price expectations is a reasonably good indicator of objective precision of expectations. Confidence is shaken by the announcement of the anticipated shock.*

Figure 15: Confidence in price expectation (NTC)

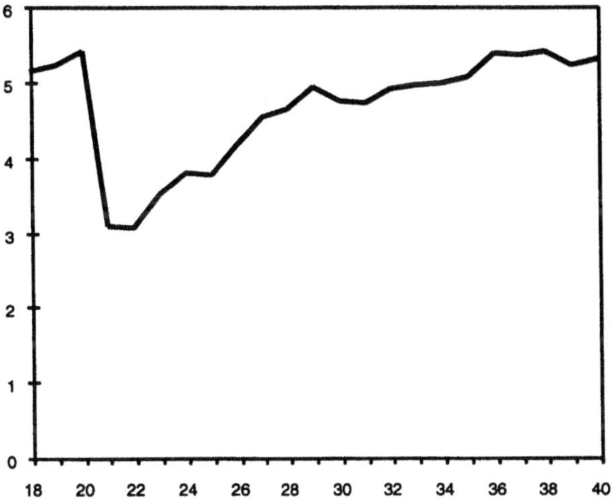

G) Summary of chapter 1

When the decision environment is characterized by strategic complementarity and represented in nominal terms, an anticipated monetary shock causes pronounced nominal rigidity (see result R1) and is *massively non-neutral* in the short run (R2). This result is found in an environment without any exogenous friction whatsoever. Thus, the New Classical hypothesis H_0 is clearly rejected. The postulate of long-run neutrality receives confirmation in our data (R3). Given their expectations, most subjects behave in a surprisingly rational way (R4). In general, subjects expect other subjects to adjust nominal prices only little to the shock (R5). These sticky expectations translate - because of strategic complementarity and best reply behavior - into sticky average prices. Sticky price adjustment translates into real income losses (R6). The loss decomposition reconfirms this conjecture: Short-run efficiency losses mainly arise because subjects do not form correct expectations after the monetary shock (R7). The monetary shock objectively and subjectively (confidence) decreases the ability to predict average prices. The announcement of the monetary seems to have fundamentally shaken expectations formation (R8).

Chapter 2: Does money illusion matter?

This chapter investigates whether money illusion is a cause of nominal rigidity and monetary non-neutrality.

Section 2.1. operationalizes the hypotheses that are to be tested.

Section 2.2. investigates the effects of money illusion given *strategic complementarity.* The behavior in three treatments is compared: The Nominal treatment (NTC), the Semi-real (SRTC) and the Real treatment (RTC). These treatments *exclusively* differ with respect to representation and are all characterized by strategic complementarity (see shaded cells in *table* 4). Overall, the money illusion hypothesis H_{A1} ("money illusion matters") is supported with strategic complements. We find that high-number illusion is at least as important as the FISHER effect.

Section 2.3. discusses the effects of nominal vs. real representation given strategic substitutes (NTS vs. RTS).

Section 2.4. summarizes the results.

Table 4: Treatments to isolate for the effects of representation

	Nominal representation	Semi-real representation	Real representation
Strategic complements	NTC	SRTC	RTC
Strategic Substitutes	NTS		RTS

2.1. Specification of hypotheses to be tested

As shown in chapter 1, money is massively non-neutral in the short run, when the environment is characterized by strategic complementarity and is represented in nominal terms (see cell NTC in *table* 4). The money illusion hypothesis H_{A1} suggests that the more nominal elements there are in the representation, the more pronounced monetary non-neutrality should be. According to this hypothesis this effect arises because people may have problems to pierce the veil of money, i.e. because they are prone to first-order money illusion. The experimental design decomposes the "veil of money" into two distinctive components (the FISHER effect and the deflationing effect, see Part II, 2.3. for details). Payoffs are represented in nominal terms in the NT, and are represented in real terms in the other two treatments. The "smallest nominal accounting unit" is held constant in the NT and the SRT, whereas it is adjusted in the RT. By representing the "veil of money" with both (NT), only one (SRT) or none (RT) of these components, we vary the ease with which economic subjects can "pierce the veil of money". Note that variations in representation constitute *ceteris paribus* variations: Subjects are confronted with *exactly the same payoff matrix in real terms* after the shock in all treatments, but this is represented differently (see *table* 1 for details). Equilibrium predictions (which are based on H_0) are of course unaffected by the variations in representation. The hypothesis H_{A1} states that "money illusion matters" because more subjects are prone to first-order money illusion, i.e. the percentage of "naive" subjects increases with more nominal elements in representation. However, H_{A1} ignores strategic considerations. This leads to the following specification of hypothesis H_{A1} (see *table* 5): The more nominal elements there are in the representation, the more pronounced observed nominal rigidity and non-neutrality. The decomposition of losses in chapter 1 has shown that the most important factor determining real efficiency losses are incorrect (sticky) price expectations. If sticky expectations in fact determine efficiency losses, and if representation affects subjects' ability to form correct expectations, expectations should be most sticky in the NT and least sticky in the RT. In addition, best reply behavior which is a measure of individual rationality, can be hypothesized to be more widespread with fewer nominal elements in representation.

Table 5: Operationalization of the money illusion hypothesis H_{A1}

Discussed in section	measure	Nominal representation	Semi-real representation	Real representation
A	Nominal rigidity	high	medium	low
B	Efficiency losses	high	medium	low
C	Percentage of best replies	low	medium	high
D	Percentage of correct expectations	low	medium	high
	Stickiness of average expectations	high	medium	low
E	Confidence	low	medium	high

2.2. The effect of nominal representation given strategic complementarity

The presentation of results proceeds as follows: Section A) discusses the effects of money illusion on nominal rigidity, section B) discusses the effects on short- and long-run non-neutrality, while sections C) to E) analyze individual-level expectations data to explain aggregate behavior. Section F) summarizes the results of this chapter. The results presented in section 2.2. have been generated by the behavior of 124 subjects who earned $29 on average. An experimental session lasted approx. 90 minutes on average.

A) Nominal rigidity

With respect to the adjustment of nominal prices the following main result emerges

> ***R9:*** *The degree of nominal rigidity is strongly and systematically affected by the nominal representation in an environment which is characterized by strategic complementarity.*

Support for this result comes from *tables* 8 and 9, and from *figure* 16. This figure shows the evolution of the average price $\bar{P}$ of the median group[11] in each of the three representations shortly before and after the shock. *Table* 7 also presents numerical values of the evolution of median and average prices over time. Since average and median prices are, in general, close to each other we do not lose much information by concentrating on median prices in the following.

Figure 16: Nominal average price of median group (NTC vs. SRTC vs. RTC)

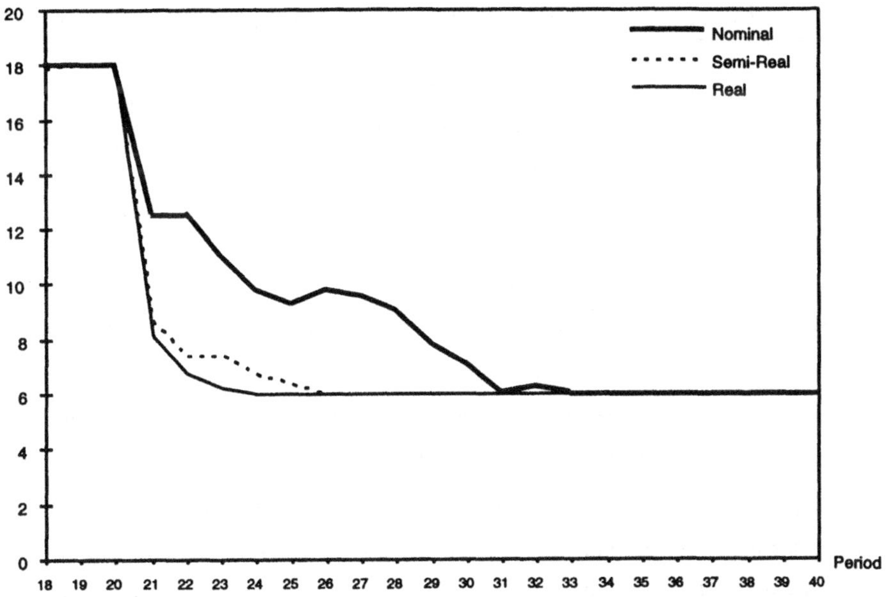

Figure 16 shows that the median group was perfectly in equilibrium ($\bar{P}^* = 18$) in each treatment before the shock while post-shock prices are strinkingly different across treatments. In the RTC we observe that the nominal price is relatively close to the equilibrium value of $\bar{P}^* = 6$ already in period 21 and that within 4 periods the equilibrium is reached. In the SRTC it takes 6 periods until equilibrium is reached while in the NTC 13 periods are needed for full adjustment. The adjustment difference is particularly large in period 21. The impact adjustment of nominal prices in the NTC is approximately only 50% of the adjustment in the other two conditions. Thus, the figure above suggests that the ranking of treatments with respect to nominal rigidity is as hypothesized (see *table* 5). As a consequence, the representation of the payoff

11. That is we calculate from the price observation of individual i in group j (P_{ij}) average group prices ($\bar{P}_j = \sum P_{ij}/n$), rank group averages $\bar{P}_j$ and show the median value. Note that the median group may change from period to period.

matrix has a *large and systematic effects* on nominal rigidity.[12]

Table 6: Percentage of price observations above, in and below the equilibrium (group averages as units of observation)

	Nominal (NTC, n =11)			Semi-Real (SRTC, n =10)			Real (RTC, n =10)		
Period	above	in	below	above	in	below	above	in	below
21 - 25	93	**7**	0	82	**16**	2	52	**34**	14
26 - 30	58	**42**	0	28	**70**	2	14	**82**	4
31 - 35	42	**56**	2	22	**74**	4	22	**78**	0
36 - 40	31	**69**	0	22	**76**	2	24	**76**	0

To provide a robust measure of price dispersion, the median absolute deviation (MAD)[13] was calculated. *Figure* 17 shows that dispersion was small in each of the treatments before the shock, since convergence of all groups was very good in each treatment. The shock increases dispersion in all treatments very similarly since all observations for period 21 are around 1.5. Remarkably, dispersion explodes from the second post-shock period on in the NTC and remains high for 10 post-shock periods. This is the case because some groups converge rather quickly, but some groups converge very slowly in the NTC. In period 23, for example, one group has already attained the post-shock equilibrium value ($\bar{P}^* = 6$), but the slowest group still has an average group price of $\bar{P} = 17.5$. In the other two treatments, group behavior is much more homogenous.

12. To test for behavioral differences across treatments non-parametric tests were conducted. Since observations within a group are not statistically independent, the average price in a group is taken as a unit of observation. A Kruskal-Wallis test reveals that there are significant differences between the three samples in the first 9 post-shock periods ($p < 0.05$, period 21-29), but not in periods 18-20 and in periods 30-40. More specifically, a median test shows that average group prices are significantly different (one tailed test, $p < 0.05$) between the NTC and RTC in the first 9 post-shock periods (periods 21-29), between the NTC and the SRTC in periods 21-24, but are not significantly different between SRTC and RTC in any post-shock period.
13. $MAD = med_i\langle |P_i - med_j(\bar{P}_j)| \rangle$, i.e. we calculate for each group i the absolute deviation of its average price from the median of all average group prices in treatment j. Then, we calculate the median of these deviations over all groups in this treatment. Note that in the pre-shock phase of the RTC we use "adjusted" prices for reasons of comparability. That is, we transform observed prices (which can be chosen in nominal steps of 3) into "deviation from equilibrium prices" $P^{RTC}(M_0) = [(P_i - P_0^*)/3] + P_0^*$. For example, an observed pre-shock average group price of 15 would be transformed into a price of $17 = [(15-18)/3] + 18$.

Figure 17: Evolution of median absolute deviation (NTC vs. SRTC vs. RTC)

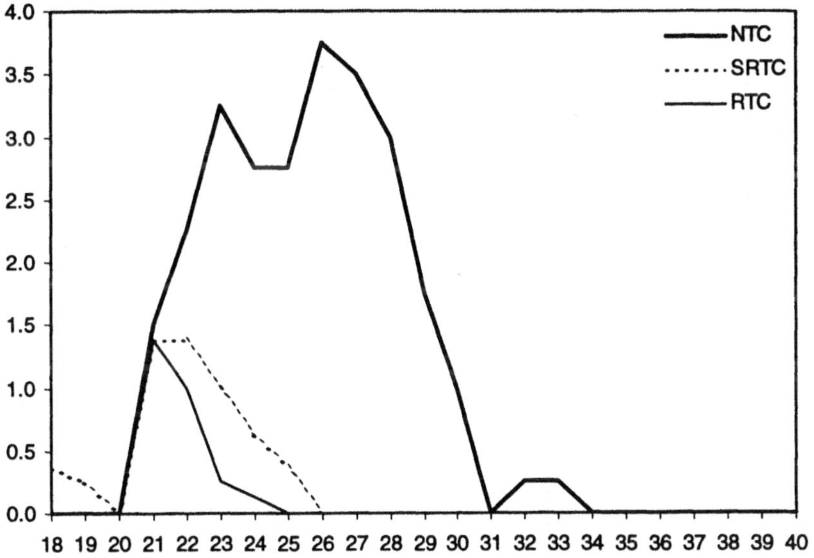

Convergence from above. *Table* 6 provides additional information about the frequencies of equilibrium play after the shock. In the NTC we observe that almost all observations (93%) are above the equilibrium and only very few (7%) are in equilibrium in the first five post-shock periods. In periods 26-30 the majority of the observations (58%) is still above the equilibrium while in the other two treatments already 70 and 82%, respectively, of the price observations are in equilibrium. *Table* 6 illustrates that *convergence is from above* in all treatments since the mass of observations was above equilibrium in the early post-shock periods and this mass is shifted to the equilibrium with differing speeds across treatments. Note how few observations below equilibrium are observed in the NTC.

Statistical analysis[14] reveals that the subjects played the equilibrium before and after the shock in the RTC, except for one single period. In the NTC, subjects played the equilibrium before the shock, but failed to reach equilibrium for 12 periods after the shock. The SRTC falls inbetween. Subjects again played equilibrium before the

14. To examine how long prices were out of equilibrium, we tested for each treatment (period by period) whether average prices are statistically different from the equilibrium value by means of a Mann-Whitney test. The hypothesis that groups are in equilibrium cannot be rejected in any treatment for the last three pre-shock periods. For the first *12, 5 and 1 periods* post-shock periods the hypothesis can be rejected in the NTC, SRTC and RTC respectively ($p < 0.05$, one-tailed)

shock, but failed to reach equilibrium for 5 periods afterwards. Thus, statistical analysis confirms that money illusion systematically affects the degree of nominal rigidity (see also *table* 5). Remarkably, we cannot reject the hypothesis that subjects play equilibrium in any treatment in periods 37 to 40. Thus, money illusion does not inhibit convergence of nominal prices in the long run.

Table 7: Evolution of nominal prices and efficiency losses over time (rounded)

	Average price of the median group			Average price			Average efficiency loss (percent)		
period	Nominal treatment (NTC)	Semi-real treatment (SRTC)	Real treatment (RTC)	Nominal treatment (NTC)	Semi-real treatment (SRTC)	Real treatment (RTC)	Nominal treatment (NTC)	Semi-real treatment (SRTC)	Real treatment (RTC)
18	18	18	18	19	18	19	3	4	3
19	18	18	18	18	18	21	1	2	12
20	18	18	18	18	18	21	1	1	9
21	13	9	8	13	9	8	65	52	32
22	13	7	7	13	8	7	47	20	11
23	11	7	6	11	7	7	35	15	5
24	10	7	6	10	7	7	27	9	6
25	9	6	6	10	7	7	17	15	5
26	10	6	6	10	7	6	16	8	4
27	10	6	6	10	6	6	16	4	1
28	9	6	6	9	6	6	11	5	1
29	8	6	6	9	6	6	9	4	1
30	7	6	6	9	7	6	14	11	2
31	6	6	6	8	6	7	8	5	10
32	6	6	6	8	6	6	6	3	6
33	6	6	6	7	6	6	6	2	3
34	6	6	6	7	6	6	5	3	0
35	6	6	6	7	6	6	3	3	0
36	6	6	6	7	6	7	10	2	10
37	6	6	6	7	6	7	5	1	3
38	6	6	6	7	6	7	14	2	4
39	6	6	6	7	6	7	12	1	12
40	6	6	6	7	6	7	2	3	7

FISHER effect vs. high-number illusion. Which one of the two components of the "veil of money" has a stronger effect on aggregate prices? It is remarkable that the "high-number illusion" (i.e. the difference between the NTC and the SRTC) is larger than the FISHER effect (i.e. the difference between SRTC and RTC)[15]. IRVING FISHER suspected that money illusion affects aggregate behavior because people fail to "perceive that the dollar, or any other unit of money expands or shrinks in value" (1928: 4). This intuition was operationalized in our experiment by holding constant the smallest nominal accounting unit in the SRT and in NTC. The only difference between the SRTC and the RTC is the smallest nominal accounting unit. When judging from the effect on nominal inertia, the difference between these treatments is relatively small (see *figure* 16) and statistically insignificant.[16] However, when judging from the effect on real efficiency losses, the relative difference between RTC and SRTC is comparable to the difference between SRTC and NTC (see *figure* 18). Yet, efficiency losses are not significantly different between SRTC and RTC.[17] We conclude from this that the FISHER effect is not the predominant reason why money illusion causes monetary non-neutrality. Rather, the high-number illusion (deflationing effect) seems to be the more important aspect of money illusion. The finding that the difference between the NTC and the SRTC is large is also remarkable from an experimental economics perspective. If one believed that *anchoring* was the driving force then real effects should have been similar in the NTC and SRTC. The reason is that the pre-shock equilibrium is in a different cell (location) of the payoff table than the post-shock equilibrium in these two treatments. Assume that subjects start to search for the "new" equilibrium (in real terms it is the same equilibrium of course) at the pre-shock prices. If this was the case, the extent of nominal rigidity should be similar in the NTC and the SRTC whereas the RTC should be the outlier. Yet, this is not the case: the SRTC is not significantly different from the RTC, whereas we find pronounced and significant differences between NTC and SRTC.

R10: *Of the two illusion creating components of the "veil of money" the high-number illusion is at least as important as the FISHER effect.*

15. In the following discussion we abstract from interaction effects which, however, are potentially important. For example, it may be the case that difference between the NTC and the SRTC is not exclusively due to high-number illusion, i.e. the deflationing effect. To test, one would have to run the RTC with nominal payoffs and compare the results to the RTC as it is.
16. Group prices are significantly different between the NTC and the SRTC in the first 4 post-shock periods, but are not significantly different in any period between the SRTC and RTC according to a one-tailed median test (see also footnote 12).
17. We find significant differences of average group profits between NTC and SRTC, but not between SRTC and RTC according to a median test (see also footnote 19).

B) Monetary non-neutrality

With respect to the question of how money illusion affects the real incomes after an anticipated monetary shock the following main result emerges:

R11: *The degree of monetary non-neutrality is massively and systematically affected by the nominal representation in an environment which is characterized by strategic complementarity.*

Figure 18 presents the evolution of efficiency losses L_{jt} for the median group in each treatment.[18] In addition, *table* 7 presents the evolution of the average value of L_{jt} over all groups in each treatment.

Figure 18: Average efficiency loss of the median group (NTC vs. SRTC vs. RTC)

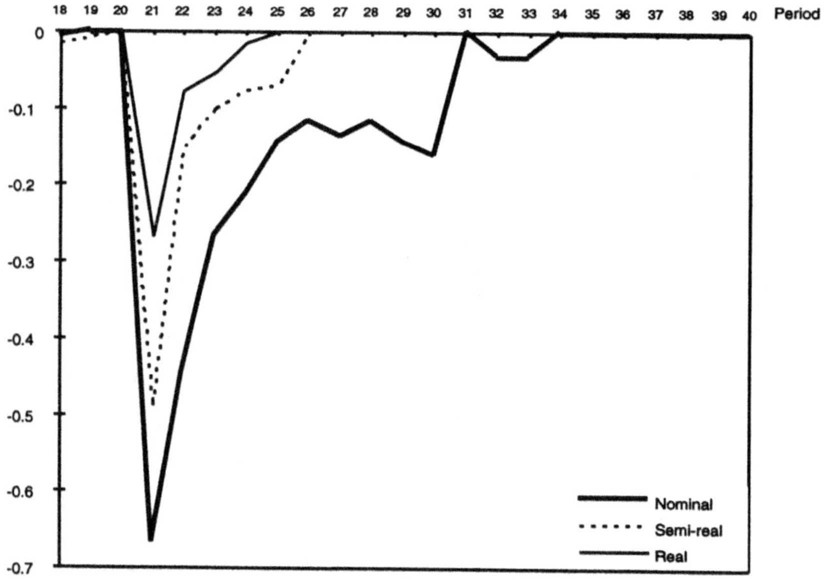

Figure 18 shows that the anticipated monetary shock is short-run non-neutral irrespective of representation. Yet, the size of the impact average efficiency loss varies considerably and systematically with representation. For example, the impact effi-

18. We first calculate by how much actual real income of group j, π_j, falls short of real income in equilibrium π^*. $L_{jt} = (\pi_{jt} - \pi^*)/\pi^*$ is a measure of the income loss of group j in period t relative to the equilibrium payoff as a percentage of the equilibrium payoff. Since the equilibrium is efficient it is also a measure of the efficiency loss.

ciency loss is more than twice as big in the NTC as in the RTC. Money illusion does not only affect the size but also the duration of efficiency losses. In line with the results on nominal rigidity the median group returns to zero efficiency losses after period 25 in the RTC and after period 26 in the SRTC while considerable losses still occur between period 25 and 30 in the NTC. In period 30, for example, the median group in the NTC still experiences an efficiency loss of more than 15%. If we aggregate the efficiency losses over $t = 21$-30, and compare them across treatments, we find that aggregate losses in the NTC are almost twice as large as those in the SRTC and almost four times as large as those in the RTC.[19] Thus, the relative magnitude of efficiency losses is as suggested by the Money illusion hypothesis (see *table* 5).

The results obtained from the *ceteris paribus* variation of representation allow to establish a truly causal relationship: Since the decision tables are exactly the same in real terms, and exclusively differ by their representation we may say that *differences* in non-neutrality are *caused* by the representation. Such large differences in efficiency losses across treatments indicate that money illusion matters, i.e. has a large economic impact.

R12: *In the long run, money is neutral irrespective of representation.*

Support for the proposition of long-run neutrality irrespective of representation comes from *figure* 16, and *tables* 7 and 6. *Figure* 16 shows that the median group is exactly in equilibrium in each treatment in periods 36-40 (see also footnote 14). *Table* 6 indicates that during these periods 69% of all group observations are *exactly* in equilibrium in the NTC while in the other two conditions even 76% of all group observations are *exactly* in equilibrium. In addition, *table* 7 shows that the deviations from equilibrium are small because *average* prices are close to equilibrium in periods 36-40. *Table* 7 also shows that the real effects of the money shock vanish over time because efficiency levels towards the end of the post-shock phase are rather similar to

19. To be precise: In total, groups in the NTC lose 26% of the potential payoff in periods $t = 21$-30. Aggregate efficiency losses in the RTC are 27% of the losses in the NTC. In the SRTC the losses are 55% of the losses in the NTC.
Statistical tests did not allow to reject the hypothesis that efficiency losses are the same in the last three periods before the shock. After the shock, we reject the hypothesis that efficiency losses are the same in the NTC and the RTC for the first ten periods (21-30) at the 5% level (Mann Whitney, one-tailed). However, according to the more restrictive median test (one-tailed, 5%-level) we find significant differences between NTC and RTC in periods 21-24, between NTC and SRTC in periods 22 and 23, and between SRTC and RTC in no period. We also find that real profits do not significantly differ across types, i.e. x-type subjects earned on average the same amount as y-type subjects in all treatments.

the efficiency levels shortly before the shock. This indicates that money is neutral in the long run. Statistical tests for differences in average prices and payoffs show that there are significant differences in the first few post-shock periods but not in later post-shock periods (see footnote 19). This indicates that money illusion has no effect on long-run non-neutrality of money.

C) Best reply behavior

With regard to best reply behavior, the following result emerges from the analysis:

R13: *Irrespective of representation a large majority of subjects plays exactly a best reply to* $\overline{P}^e_{-i}$ *in all periods and for all expectations.*

This result reconfirms our earlier conjecture of chapter 1. There, we argued that real income losses did not primarily prevail because of individually irrational behavior but rather because of strategic interaction, mediated through sticky expectations. Support for R13 comes from *figures* 19 and 20. *Figure* 19 shows the evolution of the percentage of subjects who play *exactly* best reply to their individual expectation $\overline{P}^e_{-i}$. Before the shock this percentage is between 70 and 90 percent. Immediately after the shock there is a relatively small drop in the percentage of best replies.[20]

Additional information on best reply behavior is provided by *figure* 20. This figure compares, for given intervals of price expectations $\overline{P}^e_{-i}$, the average best reply with the average level of the actually chosen nominal prices in periods 21-25.[21] The numbers above the bars indicate the relative frequency of price observations in the respective expectations intervals. In the NTC, for example, 14% of all price expectations $\overline{P}^e_{-i}$ fall within the interval 16-18 (see top diagram of the figure). *Figure 20* indicates that for any interval the deviation of actual average prices from the average best reply is relatively small in the NTC. The same holds in the SRTC and the RTC.[22]

---

20. *Figure 19* shows that the percentage of subjects choosing best replies is somewhat higher in the RTC, but differences in best reply behavior are rather small (and statistically insignificant).

21. That is, for a given period, we array all subjects by their expectation. We calculate for each subject what the best reply to this expectation would have been. The averages over all subjects in each expectation interval are calculated for the first five post-shock periods (height of white bars). The black bars show the corresponding averages of actual price choices of those subjects having a price expectation in a given interval.

The best reply for x- and y-type subjects is weakly monotonic in P_{-i}. The impression of the non-monotonicity of best replies in *figure* 20 is created by the fact that we aggregated over x and y types and that the relative frequency of x- and y-types differs across expectation intervals. Based on tests for differences in best reply behavior across types we conclude that aggregation over types is unproblematic.

Figure 19: Percentage of subjects choosing exact best replies to $\overline{P}^{e}_{-i}$ (NTC vs. SRTC vs. RTC)

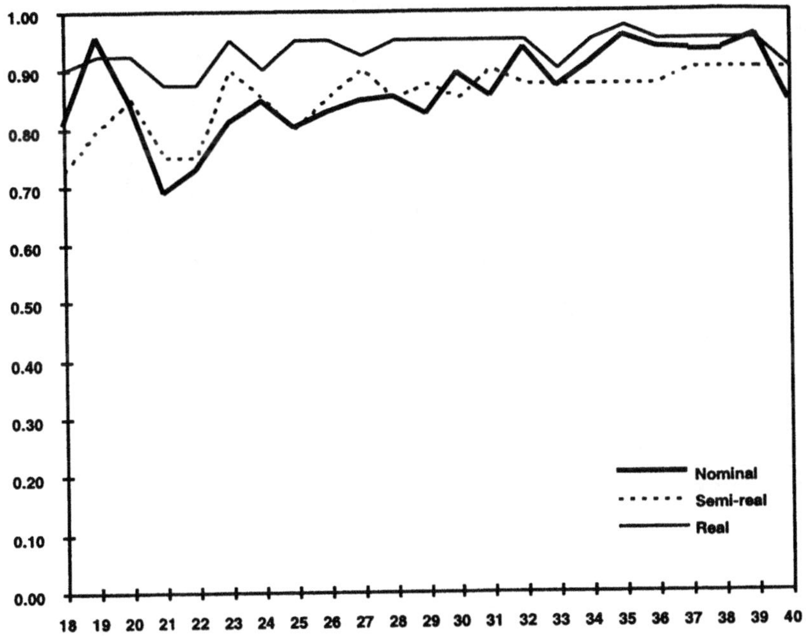

As explained in chapter 1, best reply behavior is interpreted as a measure of individual rationality. This interpretation seems reasonable since given some expectation, there is no reason whatsoever for a rational subject to deviate from best reply behavior. On the other hand, the qualitative hypothesis formulated in *table* 5 suggests that individual rationality should be lower with more nominal elements in representation. The data does not support the hypothesis that money illusion affects individual rationality in this narrowly defined sense.[23]

22. In the SRTC and the RTC occasionally relatively "large" deviations from best reply behavior are observed (e.g. in the interval 19-21 in the bottom diagram). However, these deviations are, in general, outlyers which is indicated by the small number of observations in the corresponding intervals.
23. When looking at individual data, one immediately notes that subjects overwhelmingly choose prices which are best replies to their price expectations. Yet, some outliers are observed too. However, this applies to less then 1% of all observations. The median deviation by group and period was calculated to test whether these deviations were different across treatments. The hypothesis that the median deviations from best-reply behavior are the same irrespective of representation can neither be rejected at the 5% for the three periods before the shock, nor for any single period after the shock. Thus, the median deviations from best reply-behavior are not different in the NTC and RTC.

Figure 20: Actual average prices and average best reply for given expectations (cumulative for periods 21 - 25) (NTC vs. SRTC vs. RTC)

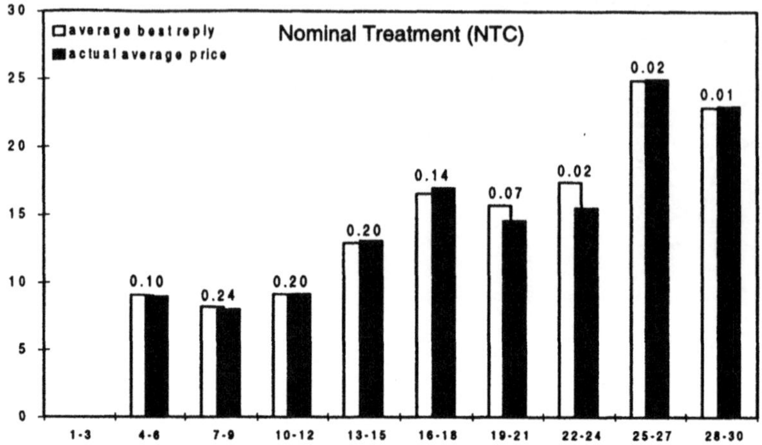

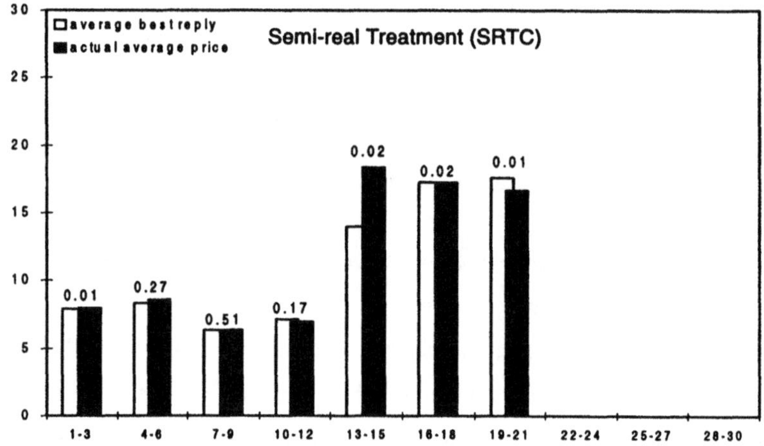

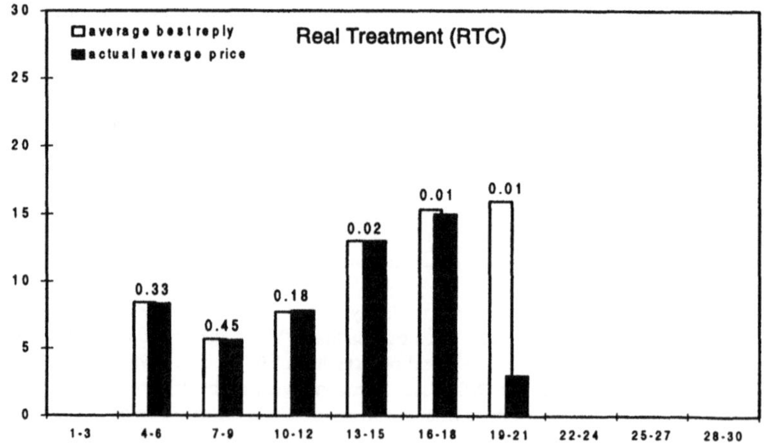

D) Price expectations $\bar{P}^e_{-i}$

The analysis in chapter 1 suggested that the main reason for real income losses after the anticipated monetary shock is to be found in sticky expectations. If this conjecture is correct, we should find that money illusion matters because representation systematically affects expectations formation. In fact, expectations are much more sticky in the NTC than in the RTC (as hypothesized in *table* 5). The first main result with respect to expectation formation is stated in

> ***R14:*** *Given strategic complements, nominal representation systematically affects price expectations. The more nominal elements in representation, the stickier price expectations.*

Figure 21: Expectation of median group on nominal average price (NTC vs. SRTC vs. RTC)

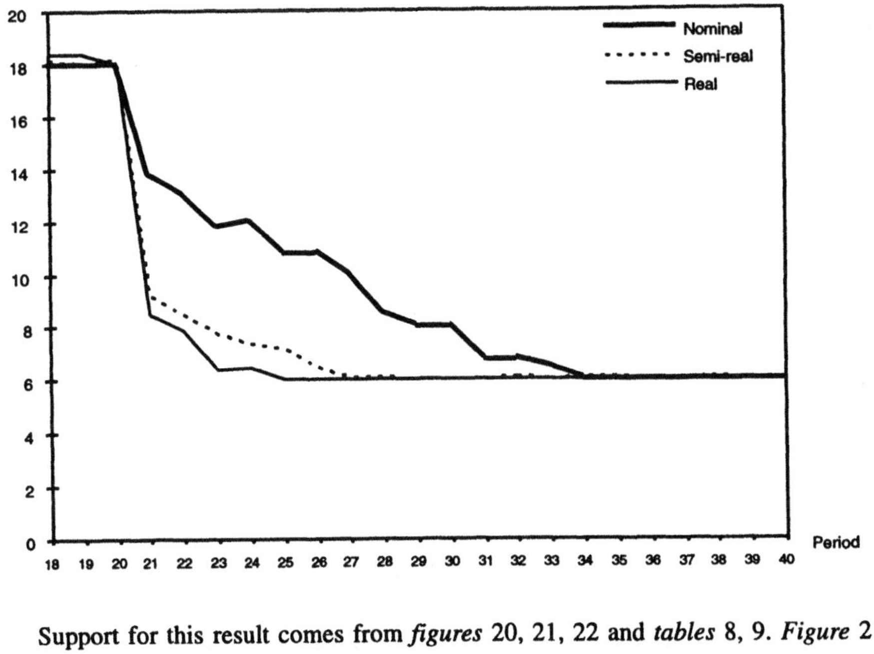

Support for this result comes from *figures* 20, 21, 22 and *tables* 8, 9. *Figure* 21 shows the evolution of the average expectation of the median group over time.[24] This picture is qualitatively strikingly similar to *figure* 16 which shows the evolution of average prices of the median group. In all three treatments price expectations exhibit some inertia but in the NTC expectations are much more sticky. The jump in price expectations immediately after the shock is more than twice as big in the RTC and the

24. The unit of observation is the average over all $\bar{P}^e_{-i}$-values in a group.

SRTC as in the NTC. Thus, subjects expect much less adjustment by other subjects in the nominal representation. This expectation is reported by subjects after having experienced an extended phase of equilibrium play and before knowing how other subjects are going to react to the anticipated shock. Moreover, while it takes 5 and 7 periods, respectively, until expectations reach the equilibrium in the RTC and the SRTC, it takes 14 periods until equilibrium expectations prevail in the NTC.

Table 8: Evolution of expectations

	Average expectation of the median group			Average expectation		
period	Nominal treatment (NTC)	Semi-real treatment (SRTC)	Real treatment (RTC)	Nominal treatment (NTC)	Semi-real treatment (SRTC)	Real treatment (RTC)
18	18	18	18	19	18	20
19	18	18	18	19	18	20
20	18	18	18	18	18	22
21	14	9	9	14	10	9
22	13	9	8	13	8	8
23	12	8	6	13	8	7
24	12	7	7	12	8	7
25	11	7	6	11	7	7
26	11	7	6	10	7	6
27	10	6	6	10	7	6
28	9	6	6	10	7	6
29	8	6	6	9	7	6
30	8	6	6	9	6	6
31	7	6	6	9	6	6
32	7	6	6	8	6	6
33	7	6	6	8	6	6
34	6	6	6	7	6	6
35	6	6	6	7	6	6
36	6	6	6	7	6	6
37	6	6	6	7	6	7
38	6	6	6	7	6	7
39	6	6	6	7	6	7
40	6	6	6	7	6	7

Table 8 reveals that the evolution of average expectations (over all groups) follows roughly the same pattern as the evolution of average expectations of the median group. Expectations exhibit some stickiness in all three treatments immediately after the shock but the stickiness is much more pronounced and adjustment takes much longer in the NTC. Information about the inertia of expectations is also provided by *figure* 20. The top diagram in *figure* 20 indicates, for example, that, in periods 21-25, 66% of price expectations in the NTC are strictly above $P^e_{-i} = 9$. In contrast, only 22% of price expectations in the RTC or the SRTC are above 9. This can be considered as rather strong evidence that expectations do not jump to the new post shock equilibrium values and that the speed with which expectations adjust, is much lower in the NTC.

Figure 22: Percentage of subjects holding correct expectations on average price (NTC vs. SRTC vs. RTC)

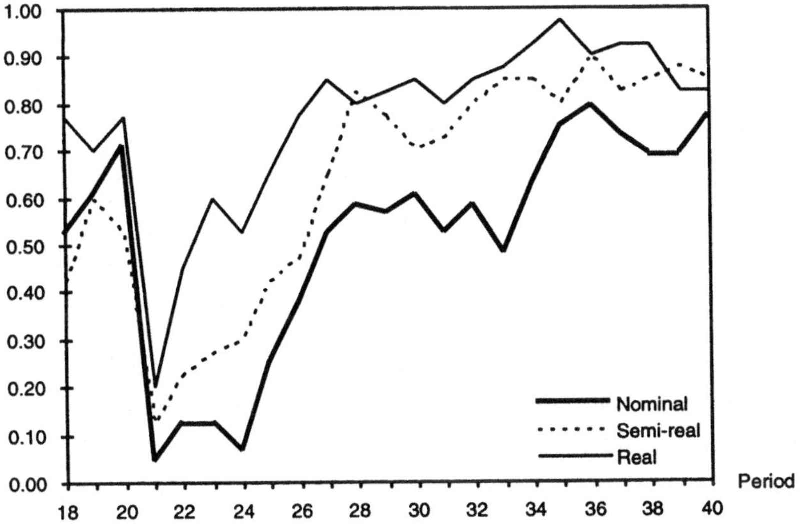

It is not only the case that subjects tended to expect relatively slow price adjustment by other subjects in the NTC and relatively quick adjustment in the RTC. They also found it much harder to predict behavior in the NTC than in the RTC and tended to expect even slower adjustment than in fact observed. *Figure* 22 shows the percentage of subjects having correct expectations on average prices. These percentages were between 70% and 80% in the RTC and the NTC in the last period before the shock. Somewhat surprisingly, the percentage of correct expectations was barely above 50% in the SRTC in period 20. After the monetary shock, the qualitative hypothesis about the ranking of this percentage (see *table* 5) is almost perfectly confirmed.

Despite the high prices that were actually chosen in the NTC, many subjects expected other subjects to choose even higher prices in the first five post-shock periods. This overestimation of group prices is more pronounced in the NTC than in the other two treatments (see *table* 9). For example, in periods 21-25 the majority of subjects (57%) expected other subjects to choose higher prices than they actually did. In contrast, the corresponding figure in the RTC is only 16%. This difference prevails despite the fact that aggregate prices are much lower in the RTC than in the NTC in these periods (see *figure* 16).

Table 9: Percentage of subjects over- and underestimating average prices (NTC vs. SRTC vs. RTC)

	Nominal (NTC, $n = 44$)			Semi-Real (SRTC, $n = 40$)			Real (RTC, $n = 40$)		
Period	over-estimation	correct expectation	under-estimation	over-estimation	correct expectation	under-estimation	over-estimation	correct expectation	under-estimation
21 - 25	57	**13**	30	49	**27**	25	16	**49**	36
26 - 30	31	**58**	11	25	**69**	7	8	**82**	11
31 - 35	32	**65**	4	15	**81**	5	6	**89**	6
36 - 40	12	**80**	8	11	**86**	4	6	**88**	7

The analysis above strongly suggests that subjects do not expect full price adjustment after a nominal shock. Moreover, as *figure* 21 shows, the results indicate that in the NTC the stickiness of price expectations is much larger than in the SRTC or the RTC. The veil of money that is incorporated in the NTC is thus responsible for much of the inertia in price expectations. In the presence of strategic complementarity and best reply behavior this inertia in price expectations causes, in turn, inertia in price choices. Therefore, in period t+1 subjects have little reason to make big adjustments in price expectations after they observed the rather small adjustment of aggregate prices in period t. The small adjustment of expectations in t+1 again provides an incentive to change prices in period t+2 only a little. Thus, small adjustments in expectations cause small adjustments in actual prices which in turn render only small adjustments in expectations reasonable.[25] The overall result of these incremental

25. Attempts to estimate expectations formation by simple adaptive expectation rules were not successful. It appears to be the case that expectation formation can neither be described by rational nor by simple adaptive processes. See also simulation results of chapter 3, shown in *figures* 30 and 31.

changes in expectations and prices is a rather slow convergence to the equilibrium.

E) Confidence in price expectation

Did subjects anticipate and perceive that their expectations became so much worse after the shock as explained above? The general pattern is that confidence[26] rises in the periods before the shock to around 5.5 in period 20. The announcement of the shock drastically reduces confidence: This is particularly true in the NTC: Confidence falls from 5.5 to 3 (= - 45%). In the RTC this change is less dramatic: Confidence falls from 5 to 3.3 (= - 35%). For the three periods directly after the shock, the relative difference in confidence amounts to 20% approx., which falls to zero by period 37 only. That is, subjects anticipate and perceive that their expectations about aggregate price behavior becomes much worse after the shock.

R15: *Given strategic complements, subjective confidence in price expectations is systematically affected as suggested by the money illusion hypothesis.*

Figure 23: Average confidence (NTC vs. SRTC vs. RTC)

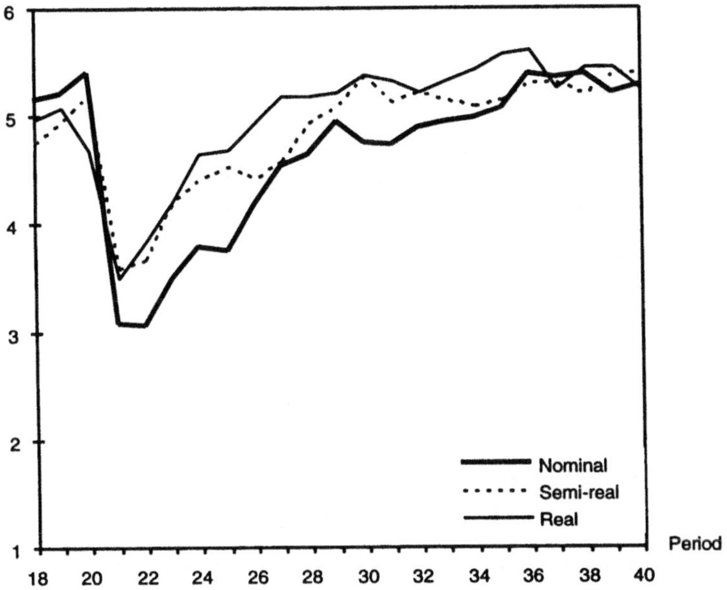

The measure of reported confidence reasonably corresponds to the actual relative imprecision across treatments (compare *figures* 22 and 23). This finding reconfirms

26. Subjects were asked to express how confident they are that the price expectation for the current period would prove to be correct by choosing an integer number from 1 (= "I am very uncertain that my price expectation will be correct") to 6 (= "I am very certain that my price expectation will be correct").

the earlier conclusion that the expectations and confidence data are of high quality. Although average confidence in the SRTC is very similar to the RTC (which should not be surprising, since the two decision tables are identical), the latter never seems to be exceeded by the former. Thus, the ranking of treatments is as suggested by the Money illusion hypothesis (see *table* 5).

2.3. The effect of representation given strategic substitutes

This section discusses the effects of a variation in the representation when the environment is characterized by strategic substitutes (NTS vs. RTS in *table* 10). It is shown that representation does not considerably affect aggregate outcomes in this case (although some differences are significant in a statistical sense). Thus, hypothesis H_{A1} appears to be incomplete, whereas the "Money illusion *cum* strategic anticipation" hypothesis H_{A3} receives support by the data. The following discussion is very concise since chapter 3 discusses the effects of strategic complements vs. substitutes in detail.

Table 10: Overview over treatments

	Nominal representation	Semi-real representation	Real representation
Strategic complements	NTC	SRTC	RTC
Strategic Substitutes	NTS	-	RTS

The results presented in this section have been generated by the behavior of 92 subjects (44 in the RTS and 48 in the NTS). The presentation of results proceeds as follows: The effects of nominal vs. real representation on nominal rigidity are discussed in section A), and the effects on monetary non-neutrality are discussed in section B).

A) Nominal rigidity

The main result with respect to nominal rigidity is summarized in

R16: *Given strategic substitutes, nominal vs. real representation does not cause significant nominal inertia.*

Figure 24: Average price of median group (NTS vs. RTS)

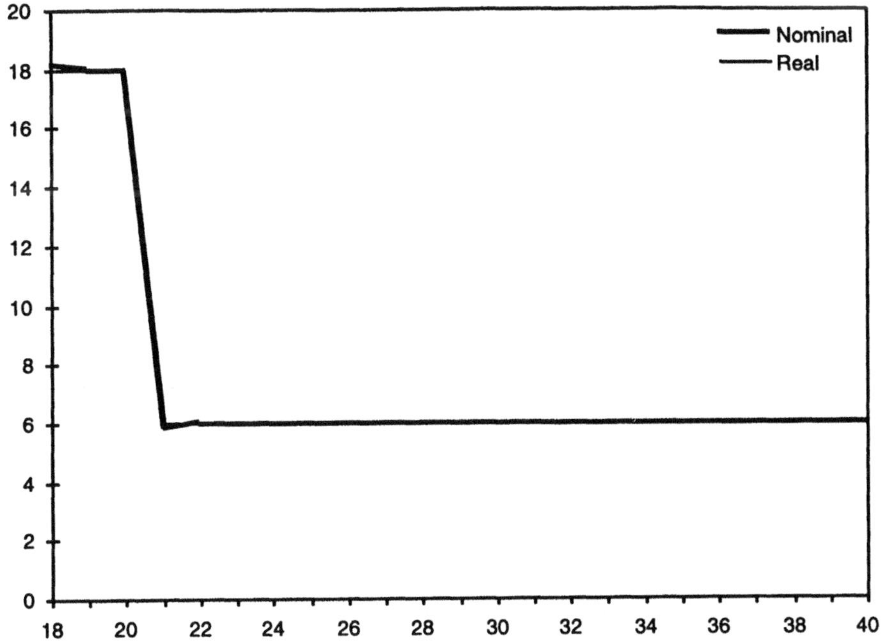

Support for this result comes from *figure* 24. This figure shows that there is no nominal inertia at all, irrespective of representation. Given strategic substitutes, nominal average prices of the median group literally jump to the new equilibrium levels in both representations. The hypothesis that average group prices are the same cannot be rejected in any post-shock period according to a median test (one-tailed, 5%-level). This finding contrasts with the results from the effect of representation when the environment is characterized by strategic complementarity (NTC vs. RTC). In this case, statistical tests yielded the following result: the hypothesis that representation does not matter can be rejected for *10 periods* after the shock. Hence money illusion causes nominal inertia in the complements, but not in the substitutes treatment (compare *figures* 16 and 24). We conclude for the moment that money illusion does not affect nominal rigidity as hypothesized in H_{A1} (see *table* 5). However, the differences between strategic complements and substitutes will be discussed in much more detail in chapter 3.

B) Monetary non-neutrality

The main finding with respect to monetary non-neutrality is summarized in

R17: *An anticipated monetary shock is largely short-run neutral in the RTS. Money illusion has a slight (though significant) effect on non-neutrality.*

A comparison of efficiency losses of the median group in the NTS and the RTS (see *figure* 25) shows that efficiency losses are very small in both treatments. *Money is largely short-run neutral* in *both* treatments. Thus, the representation did not have a considerable impact on real incomes when the environment is characterized by strategic substitutes. Yet, differences are statistically significant according to a median test (one-tailed, 5%-level). Results show that average profits are only significantly different in period 21 with strategic substitutes (NTS vs. RTS). Hence, the "simple" (non-interactive) Money illusion hypothesis is clearly rejected with respect to nominal rigidity but not with respect to monetary non-neutrality. It should be noted that the effect of money illusion is much more pronounced with strategic complements than with substitutes (compare *figures* 18 and 25). Nonparametric tests reconfirm this optical impression: Whereas average group profits were significantly different between NTC and RTC for 4 post-shock periods (see footnote 19), they are only different in the first post-shock period between NTS and RTS. This finding does not allow to reject the Money illusion hypothesis H_{A1} altogether. Yet, the hypothesis that "money illusion matters more when strategic complementarity prevails" (H_{A3}) organizes the data very well. This hypothesis will be analyzed in much more detail in chapter 3. For now, we summarize our findings in

R18: *Money illusion causes more pronounced monetary non-neutrality with strategic complements than with strategic substitutes.*

Figure 25: Average efficiency loss of the median group (NTS vs. RTS)

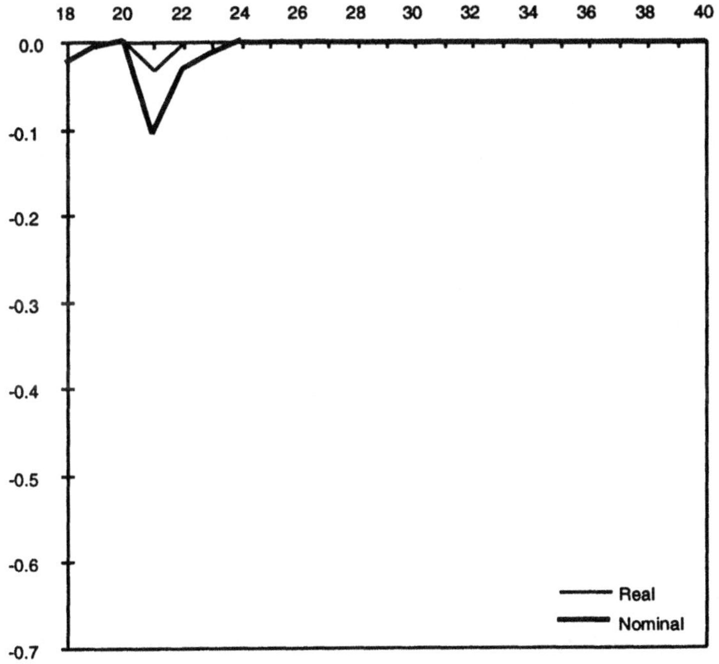

2.4. Summary of chapter 2

Section 2.1. operationalized hypothesis H_{A1}. According to this hypothesis money illusion directly causes aggregate effects because subjects are prone to first-order money illusion.

Section 2.2. has shown that money illusion matters, i.e. causes pronounced nominal rigidity (see result R9) and considerable monetary non-neutrality (R11) after an anticipated monetary shock, if strategic complementarity prevails. Yet, money illusion cannot prevent money to be neutral in the long run (R12). Of the two illusion creating effects, the high-number illusion seems to be the more important than the FISHER effect (R10). A large majority of subjects chose best replies given their expectation, irrespective of representation. Money illusion thus did not affect individual rationality in this narrowly defined sense (R13). The Money illusion hypothesis H_{A1} is supported with respect to the stickiness and correctness of expectations. When the veil of money is "less transparent", subjects expect other subjects to choose stickier prices (R14) and find it more difficult to correctly predict average prices. As a consequence, subjects' confidence in their predictions is systematically lower with more nominal elements in representation (R15).

In all, money illusion is a causal factor explaining the degree of nominal inertia and monetary non-neutrality. However, the exact mechanism by which money illusion affects the aggregate remains unclear. The data do not allow to exactly discern between the direct effects of first-order money illusion, the (correct) anticipation of first-order money illusion, and the (incorrect) belief that other subjects are prone to first-order money illusion. However, the nominal vs. real representation affects subjects' ability to predict the others' behavior. Representation seems to affect one's notion of how rational *others* are. Thus, the results provide some support for the notion that representation systematically affects "higher-order money illusion".

Section 2.3. has shown that the effect of representation on nominal rigidity (R16) and monetary non-neutrality (R17) are negligible when the environment is characterized by strategic substitutes. The results clearly show that money illusion matters more with strategic complements than with substitutes (R18).

In all, money illusion matters, but it is not (yet) clear why. In particular, the hypothesis H_{A1} which ignores strategic aspects appears to be incomplete. For the moment, we conclude that the *interaction* of the strategic environment with money illusion is important and that H_{A3} ("money illusion matters more with strategic complements than with substitutes") is supported. The next chapter will shed more light on the question why this is the case.

Chapter 3: The effects of strategic complements and strategic substitutes

This chapter investigates whether strategic complementarity is a cause of nominal rigidity and monetary non-neutrality.

Section 3.1. deduces and operationalizes the hypotheses to be tested.

Section 3.2. discusses the effects of varying the strategic property in the *nominal* representation (NTC vs. NTS in *table* 11). It is shown that the pronounced nominal rigidity and massive non-neutrality observed with strategic complements almost completely vanishes when the environment is characterized by strategic substitutes. Thus, the HALTIWANGER / WALDMAN hypothesis H_{A2} appears to be supported when the environment is represented in nominal terms.

Section 3.3. investigates the effects of strategic substitutes and complements when the decision environment is represented in *real* terms (RTC vs. RTS). In this case, the behavioral effects of complements vs. substitutes are rather small. Thus, hypothesis H_{A2} is incomplete, whereas the "Money illusion *cum* strategic anticipation" hypothesis H_{A3} is supported.

Table 11: Treatments to isolate for the effect of strategic complements vs. strategic substitutes

	Nominal representation	Semi-real representation	Real representation
Strategic complements	NTC	SRTC	RTC
Strategic Substitutes	NTS	-	RTS

3.1. Specification of hypotheses to be tested

As explained in Part I (see page 41), hypothesis H_{A2} states that "strategic properties matter" because super-rational agents magnify the effects of naive subjects with strategic complements or mitigate them with strategic substitutes. The hypothesis remains silent on how money illusion affects the proportions of these agents or their respective expectations. The operationalization of H_{A2} is explained in section A). Section B) discusses the operationalization of hypothesis H_{A3} which states that strategic properties and money illusion interact, i.e. that strategic properties matter more with money illusion, and money illusion matters more with strategic complements.

A) Specification of hypothesis H_{A2}

The HALTIWANGER / WALDMAN (HW) theory assumes that agents are heterogeneous with respect to rationality. In particular, these authors assume that there are super-rational agents which have perfect foresight, and (extremely) naive agents which hold fully adaptive expectations and choose best replies to these expectations. Since the HW theory remains silent on why naive agents should behave irrationally, the predictions of H_{A2} are independent of the nominal vs. real representation. The specification of hypothesis H_{A2} is based on simulations (see *figures* 30 and 31). These simulations were run under the assumption that some agents have fully adaptive expectations, whereas the remaining agents have perfect foresight. The exact predictions of H_{A2}, of course, depend on the proportions of these two types of agents. Some general properties can nevertheless be deduced from intermediate cases (the proportion of super-rational agents varies between 0% and 75%[1]): In the CT, nominal prices should converge from above, whereas they should first overshoot and then oscillate in the ST (see *table* 12). The deviations from equilibrium (i.e. the degree of stickiness in the CT and the extent of overshooting in the ST) are the more pronounced the larger the proportion of naive agents. Simulations show that real efficiency losses last longer in the CT than in the ST, but are more pronounced in the ST than in the CT. The differences in aggregate behavior result from different reactions of the super-rational agents to their *correct expectations* about the behavior of the naive ones. The latter hold fully adaptive expectations, and are assumed to play best reply to these expectations independently of strategic properties. Hence, the percentages of best replies and correct expectations, as well as confidence should be the same in the CT as in the ST. However, average impact expectations (in period 21) should in general be different in the CT and the ST. Super-rational agents should expect prices (far) above equilibrium in the CT but closer to the equilibrium in the ST. As a consequence, average expecta-

1. With 100% super-rational agents, the prediction of H_{A2} obviously degenerates to the Neutrality case H_0.

tions should be more sticky in the CT than in the ST.

Table 12: Specification of hypothesis H_{A2}

Discussed in section	measure	Strategic complements	Strategic substitutes
A	Nominal rigidity ($\bar{P}$)	pronounced rigidity	no rigidity
B	Efficiency losses $(\pi - \pi^*)/\pi^*$	long-lasting	more pronounced impact effect than CT, short duration
C	Percentage of best replies	same as substitutes	same as complements
D	Percentage of correct expectations	same as substitutes	same as complements
D	Stickiness of average expectations in period 21 ($\bar{P}^e$)	pronounced	less pronounced than CT
E	Confidence	same as substitutes	same as complements

B) Specification of hypothesis H_{A3}

The hypothesis H_{A2} combines the "simple" (non-interactive) money illusion hypothesis with hypothesis H_{A3} by postulating that money illusion and strategic properties interact in a particular way. As in H_{A1}, naive agents are assumed to be subject to first-order money illusion which by assumption leads them to choose prices which are not best replies. As in H_{A2}, the sophisticated agents are assumed to be free of first-order illusion and to anticipate that others may be prone to money illusion. In combination, H_{A3} states that the more subjects are expected to be naive the more nominal elements there are in representation. Because both aspects (the FISHER effect and the high-number illusion) of money illusion draw the behavior of the (supposedly) naive to high prices in the NTC, expectations and average prices should be more sticky in period 21 in the NTC than in the SRTC and in the RTC. In the NTS, first-order money illusion draws behavior to high or low prices, depending on which aspect of money illusion dominates. Given that both the FISHER effect and the high-number illusion have some behavioral relevance, behavior of naive agents should be relatively close to equilibrium. Hence, the ranking with respect to stickiness of expectations according to H_{A3} is NTC > SRTC > RTC, NTC > NTS, RTC > RTS (see *table* 13). If best reply behavior is the same in all treatments, these sticky expectations translate into sticky prices in the CT (NTC > SRTC > RTC) but into overshooting prices in the ST (NTS < RTS). The interaction of higher-order money illusion and strategic properties comes in because subjects are hypothesized to expect more inertia in the NTC than in the NTS, but expect similar inertia (since there is no high-number illusion in this case) in the RTC and the RTS. Hence, the difference in inertia and real effects NTC - NTS is larger than RTC - RTS, i.e. strategic properties matter more with money illusion. Moreover, it is hypothesized that money illusion matters more in the CT than in the ST because of high-number illusion. Nominal payoffs are highest at prices far above the equilibrium in the NTC, but highest nominal payoffs are at low prices (relatively close to the equilibrium) in the NTS (see payoff tables in appendix A2).

Table 13: Specification of hypothesis H_{A3}
(the earlier the letter appears in the alphabet, the more pronounced the effect)

Section	measure	NTC	SRTC	RTC	NTS	RTS
A	Nominal rigidity	A	B	C	E	D
B	Efficiency losses	A	B	C	a (< A)	b (< a)
C	Percentage of best replies	equal	equal	equal	equal	equal
D	Percentage of correct expectations	C	B	A	C	A
	Stickiness of expectations ($t = 21$)	A	B	C	a (< A)	b (< a)
E	Confidence	C	B	A	C	A

3.2. The effect of strategic substitutes vs. complements when the environment is represented in nominal terms

The presentation of results in this section proceeds as follows: Section A) discusses the effects of the strategic properties on nominal rigidity, section B) discusses the real effects of the monetary shock. Section C) discusses best reply behavior, sections D) and E) price expectations.

A) Nominal rigidity

The main result with respect to nominal rigidity is stated in

R19: *Nominal rigidity is much more pronounced and lasts much longer with strategic complements than with strategic substitutes, given the nominal representation.*

Support for R19 comes from *table* 14 and *figure* 26. This *figure* shows the evolution of nominal average prices in the two treatments. In contrast to the NTC where we observe massive nominal inertia in the aggregate (see chapter 1), nominal prices literally jump to the new equilibrium level in the NTS. Hence, we observe a *massive treatment effect*: the controlled variation of the strategic property from complements to substitutes reduces nominal rigidity to almost zero. Thus, the Money illusion hypothesis H_{A1} which states that the aggregate behavior should be the same in both treatments (see *table* 5) is rejected.[2] Instead, the hypothesis H_{A2} is supported (see *table* 12).

2. A median test of the null hypothesis that average prices are the same in both treatments (treating a group as an observation) yields the following results: before the shock (periods 18-20) we cannot reject this hypothesis at the 5%-level. After the shock, we can reject this hypothesis for *12 periods* (per. 21-28 and 30-33, 5%-level, one-tailed).

Figure 26: Evolution of nominal average price of median group (NTC vs. NTS)

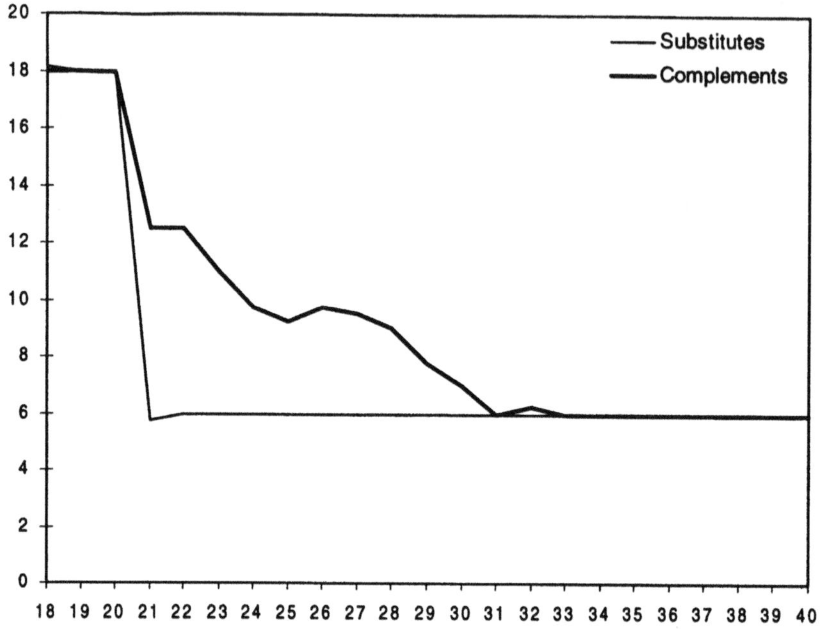

Table 14 shows the percentage of subjects who choose prices above, in or below equilibrium. In the last three periods before the shock (per. 18-20), approximately 80% of subjects choose equilibrium prices, with the rest of subjects split evenly above and below equilibrium in both treatments. In the NTS, the percentage of subjects choosing equilibrium prices remains at high levels and the *distribution of choices remains largely unaffected* by the monetary shock. The NTC imparts a completely different impression. Before the shock, the percentage of subjects who choose equilibrium prices is comparable to the case with substitutes. In the post-shock periods this percentage falls dramatically. In addition, the price distribution becomes extremely *skewed.* For example, 84% of subjects choose prices that are above equilibrium while hardly anyone chooses prices below equilibrium in periods 21-25. This skewness of the price distribution reflects the large degree of nominal rigidity observed in this treatment.

In chapter 2 we already noted that convergence is from above in the NTC. In the NTS, the percentage of equilibrium choices remains almost unaffected and convergence, if anything, is from below. Note, for example, that nominal prices slightly overshoot in period 21 in the NTS (see *figure* 26).

Table 14: Percentage of observations above, in and below the equilibrium (Subjects as units of observation)

	Complements (NTC, $n = 44$)			Substitutes (NTS, $n = 48$)		
Period	above	in	below	above	in	below
18 - 20	14	**83**	3	13	**77**	10
21 - 25	84	**16**	0	9	**77**	14
26 - 30	55	**45**	0	6	**84**	10
31 - 35	33	**67**	0	3	**91**	5
36 - 40	18	**82**	0	4	**91**	5

Individual price choices. Closer examination of the first period following the monetary shock reveals how dramatic the effect of a variation of the strategic property of the decision environment on nominal rigidity is. *Figure* 27 shows the distribution of nominal prices in the period before and after the shock in the two treatments (for subjects of type y). The two diagrams in the upper half show that most subjects choose equilibrium prices (= 27) before the shock in both treatments. After the shock, the distribution of nominal prices is dramatically different in the two treatments (see lower half of the figure): with strategic substitutes we observe that most (70%) subjects choose again equilibrium nominal prices (= 9) immediately after the shock. In the case of strategic complements this is true only for a small minority of subjects (20%). Another feature of the two lower diagrams is even more striking: while we observe massive nominal inertia with strategic complements (most price decisions are far above equilibrium), no such inertia can be observed with strategic substitutes. Here, we observe only one subject who chooses a nominal price above equilibrium. On the other hand, we observe a lot of "overshooting" in the NTS: 25% of subjects choose nominal prices which are below equilibrium, where no such overshooting is observed in the NTC. Sticky price choices in the NTC and overshooting in the NTS can be explained by strategic anticipation, i.e. by "sophisticated" subjects who expect other "naive" subjects not to respond to the shock, e.g. because they expect them to be nominally anchored at the pre-shock equilibrium prices.

Figure 27: Distribution of nominal prices (NTC vs. NTS, for type *y*)

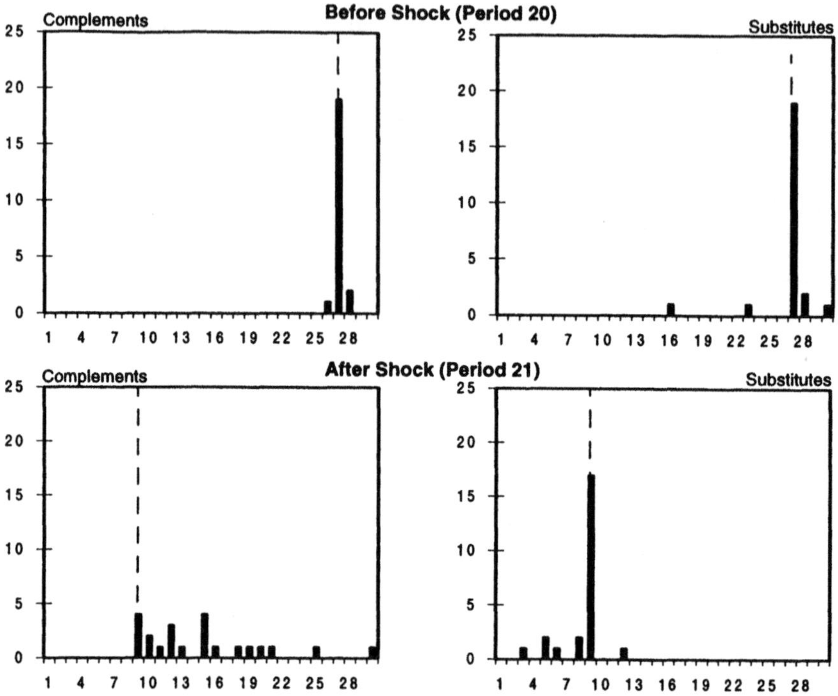

Hence, behavior can be rationalized by the HALTIWANGER / WALDMAN hypothesis H_{A2}. Alternatively, one could explain observed price behavior by first-order high-number illusion. As will be discussed in section D), the variation of the strategic properties also affected (impact) expectations.

B) Monetary non-neutrality

With respect to monetary non-neutrality the following result emerges:

R20: *Given the nominal representation, the anticipated monetary shock is associated with considerable real income losses. These losses are much more pronounced and last much longer in the complements treatment than in the substitutes treatment.*

Support for R20 comes from statistical analysis and from *figure* 28. This figure shows efficiency losses of the median group in the NTC and NTS. After the shock efficiency losses of the median group are much more pronounced and last much longer in the NTC than in the NTS. For example, average impact efficiency losses are 65% in the NTC but only 18% in the NTS. The differences are even more pronounced when we compare the respective values for the median groups (see *figure* 28). Real income losses are still substantial from periods 25-30 in the NTC, whereas the median group returns to zero real income loss from period 24 on in the NTS.[3] Thus, strategic properties matter much for short-run non-neutrality when money illusion can be supposed to be prevalent. On the other hand, strategic properties do not seem to matter in the long run. In all, the HALTIWANGER / WALDMAN hypothesis H_{A2} receives partial empirical support judging from monetary non-neutrality. According to the hypothesis, impact non-neutrality should be more pronounced in the NTS (see *table* 12 and *figure* 31) which is obviously not the case.

Figure 28: Real efficiency loss of the median group (NTC vs. NTS)

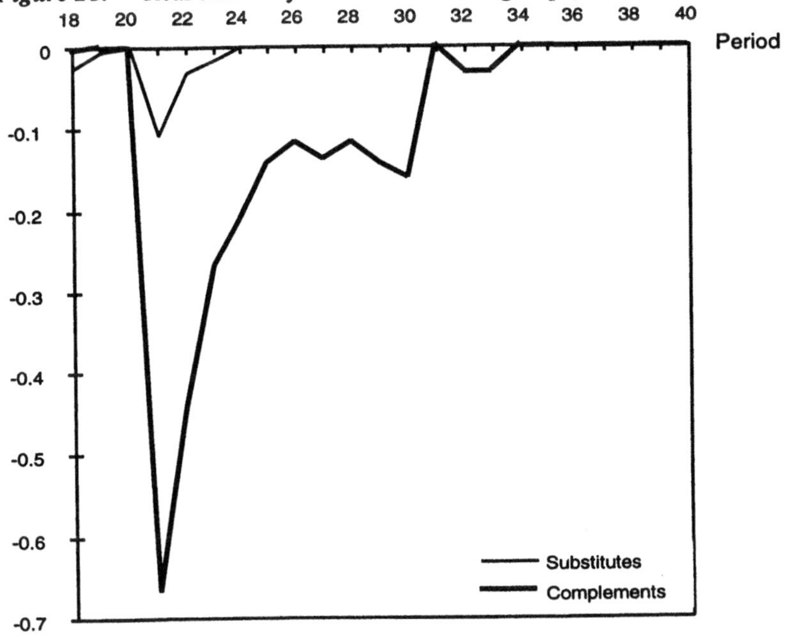

C) Best reply behavior

With regard to best reply behavior we find the same behavior as in all other treatments:

> ***R21:*** *In all periods and for all ranges of expectations a large majority of subjects chooses an exact best reply to P_{-i} .*

3. The hypothesis was tested that real profits by group and period are the same in the NTC and in the NTS. This hypothesis cannot be rejected for the last three pre-shock periods. It is rejected for periods 21-24 (4 periods) after the shock (also in period 32). In all other periods (25-31, 33-40) it cannot be rejected.

Support for R21 is provided by *figure* 29. This figure shows the percentage of subjects who choose prices which were best replies given their price expectation. In general, the percentages are quite high in both treatments. However, the variation in the strategic property seems to somewhat affect individual rationality.[4]

Figure 29: Percentage of subjects choosing best replies (NTC vs. NTS)

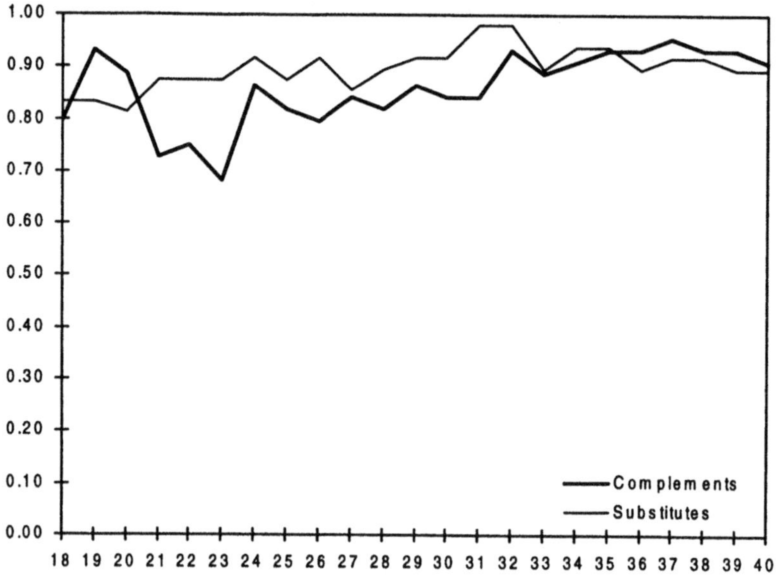

Simulated vs. actual price adjustment. Note that the HALTIWANGER / WALDMAN hypothesis states that for *given* sticky behavior of naive agents, super-rational agents should choose low prices in the NTS but relatively high prices in the NTC. We in fact observe sticky aggregate prices in the NTC and no inertia of aggregate prices in the NTS (see *figure* 26). The hypothesis suggests that if subjects expect the same price, behavior will be different in the two treatments. The respective theoretical predictions are illustrated in *figures* 31 and 32). These simulations show the adjustment process of nominal prices and efficiency losses for one, two, three or four quarters of fully adaptive ("naive") subjects.[5] Naive agents are simulated as having fully adaptive expectations. The super-rational subjects are simulated as having perfect foresight. Hence, as

4. An Exact Fisher test for all periods under consideration reveals that there are no significant differences in the percentages of best responders at the 5% level, except for periods 21 ($p = 0.04$) and 23 ($p = 0.02$).
5. See HALTIWANGER and WALDMAN (1989) for similar simulations of unanticipated shocks.

suggested by result R21 all players are assumed to choose a best reply to their expectation. The difference between the two types of players is that the naive choose best replies to last periods' realization, and the super-rational choose best replies to this period's actual average price. If, for example, all agents are naive, each expects an average price of $\overline{P}_{-i}^{e}$ = 18 in period 21. The average best reply to this expectation is $\overline{P}$ = 17 in the CT, but $\overline{P}$ = 1 in the ST (see heavy lines in *figure* 30). If 75% of agents are naive, average prices far above ($\overline{P}$ = 11) result in the CT, but a price close to equilibrium ($\overline{P}$ = 6) prevails in the ST (see thin dotted lines in *figure* 30). This price adjustment translates into monetary non-neutrality as shown in *figure* 31.

Figure 30: Simulated adjustment of average price with different shares of adaptive players: complements vs. substitutes

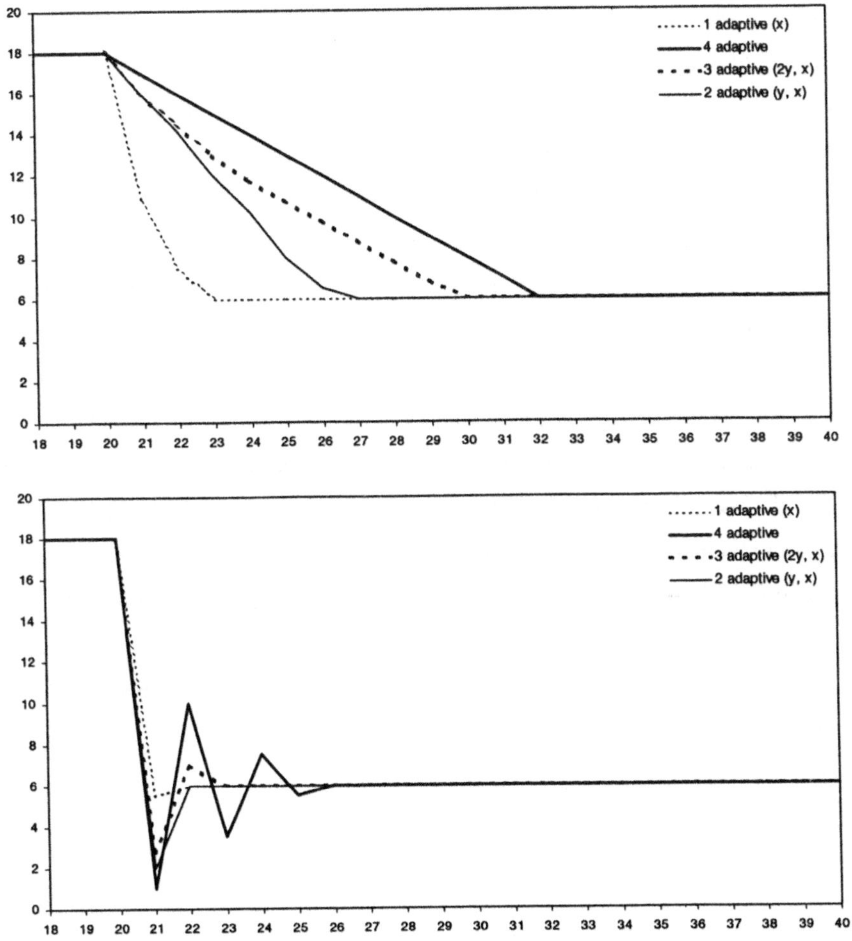

The general finding from these simulations is that disequilibrium prices and efficiency losses are the more pronounced the larger the proportion of adaptive players. In the CT, nominal rigidity is more pronounced with more naive players. In the ST, nominal prices tend to overshoot more with larger proportions of naive players. Due to our parametrization, increasing the proportion of naive (adaptive) players markedly increases the impact effect on efficiency in the substitutes treatment, whereas it also increases the medium term effects on efficiency in the complements treatment. Note that the scales are different in the two diagrams of *figure* 31. The simulated impact efficiency losses are considerably larger in the substitutes treatment.

How does the evolution of actual prices compare with the simulated price series (compare *figures* 26 and 30)? In the substitutes treatment, the simulation with one fourth of adaptive agents replicates the price series quite well. In the complements treatment neither of the simulations replicates the actual adjustment of prices. In particular, nominal prices exhibit a larger impact adjustment in the NTC in period 21 than in any of the simulations. This suggests that subjects were not fully adaptive. On the other hand, price adjustment after period 21 is even slower in the NTC than in the simulation with four adaptive players.

When comparing actual and simulated efficiency losses (compare *figures* 28 and 31), we again find that the simulation with one quarter of adaptive agents reasonably replicates efficiency losses in the NTS. In the NTC, actual efficiency losses are best replicated by the simulation with four adaptive agents. However, actual impact efficiency losses are somewhat larger in the NTC than in the respective simulation.

What can we learn from this comparison? First, the behavior of experimental subjects conforms to the prediction of H_{A2} (i.e. the simulation) insofar as prices are more sticky and real income losses are more pronounced in the NTC than in the NTS. Second, aggregate behavior in the NTS looks as if subjects were "more rational" than in the NTC. The observed series is (given the assumptions of H_{A2}) best replicated by assuming that 25% of subjects are naive in the ST but one has to assume that all agents are naive in the CT. The section D) provides some explanations for this puzzling observation.

Figure 31: Simulated average efficiency loss with different shares of naive (fully adaptive) players: complements vs. substitutes

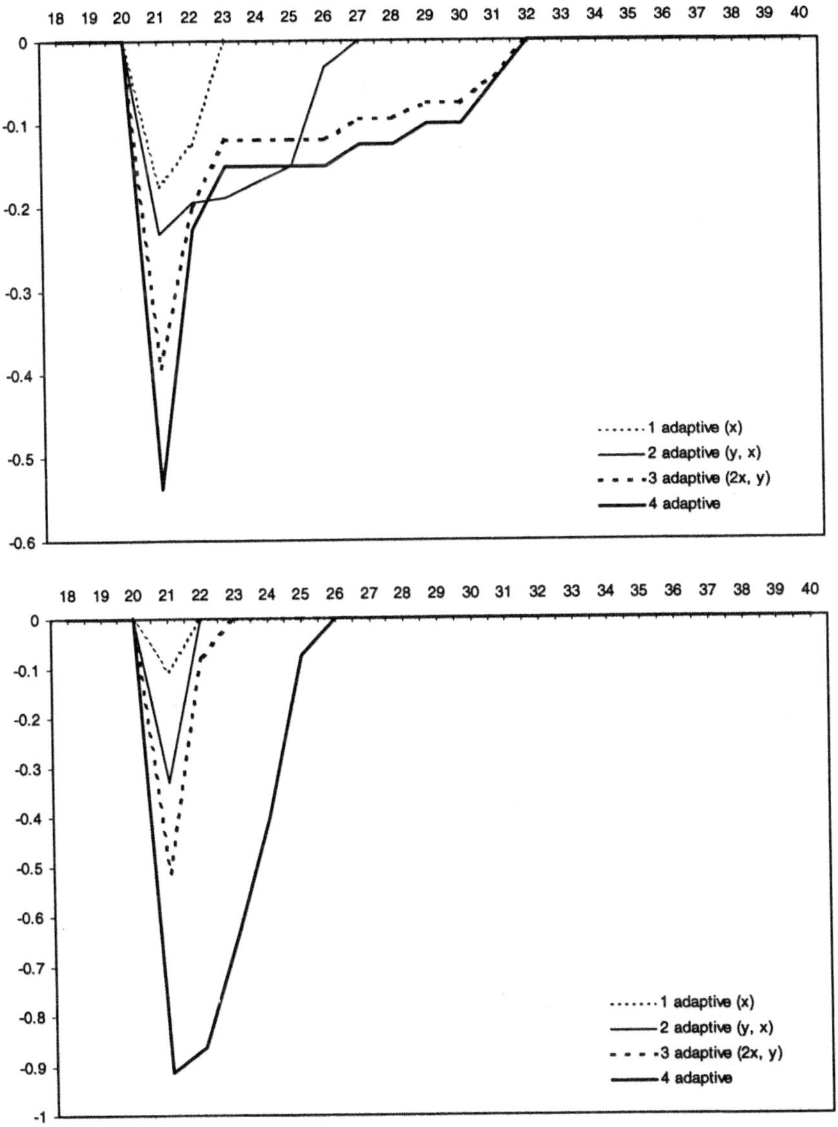

D) Price expectations $\bar{P}^{e}_{-i}$

Strategic properties affected expectations formation since *expectations were markedly different* in the two treatments. Obviously, the strategic properties seem to affect the ability of agents to form correct impact expectations and to learn from post-shock observations. In general, it seems to be more difficult to learn in the complements treatment than in the substitutes treatment. This finding is not implied by the HALTI-

WANGER / WALDMAN hypothesis.

Figure 32 shows that impact expectations are much closer to the equilibrium expectation (= 6) in the NTS than in the NTC. It seems to be the case that it is cognitively simpler to form equilibrium expectations in the NTS. As will be explained below, expectations were much more correct in the NTS than in the NTC. Despite the fact that the NTS and NTC have been generated by a *ceteris paribus* variation of the functional form, the two payoff tables may not be cognitively equivalent.

Figure 32: Expectations and average prices (NTC vs. NTS)

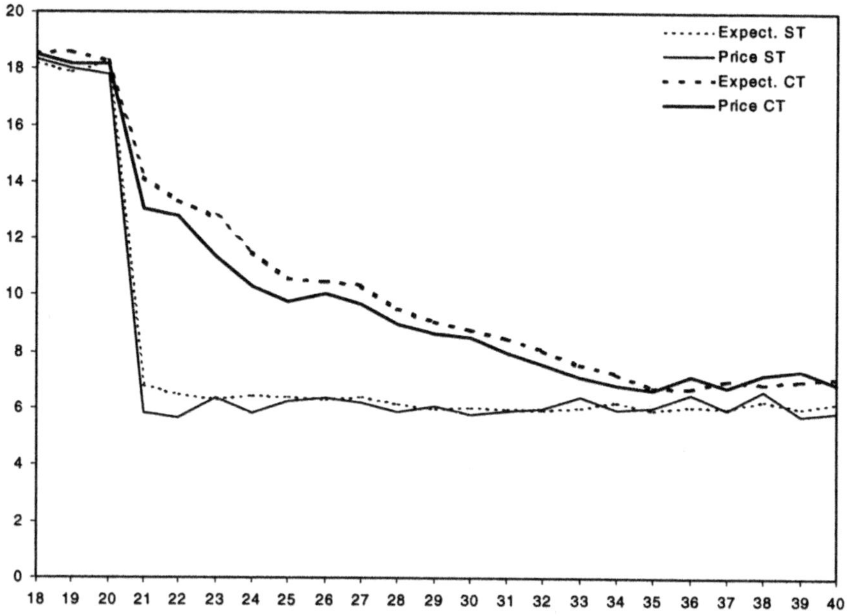

High-number illusion as an explanation for differing expectations. Nominal anchoring and high-number illusion draws the behavior of naive and sophisticated agents to high prices in the NTC. In contrast, (the anticipation of) money illusion may draw behavior either to high or low prices. If anticipating agents expect some mixture of the two aspects of money illusion to prevail, they tend to choose prices close to the equilibrium in the NTS, but prices far above the equilibrium in the NTC. Paradoxically, the anticipation of money illusion may explain the high incidence of equilibrium play in the NTS (see *figure* 27).

Cognitive explanation for effect of strategic properties on expectations. Alternatively to the explanation given above, a psychological reason may explain why it is easier to form equilibrium expectations in the NTS than in the NTC: It may be the case that the "*degree of consistency in reasoning*" is psychologically larger in the NTC than in the NTS. To illustrate, suppose that subjects were somehow anchored in the nominal pre-shock equilibrium price (since a majority subjects chose this nominal prices throughout most pre-shock periods). A subject in the NTC may reason the following way: "Suppose everybody else chooses the same price as before. What should I do? I should choose a price which is slightly lower than the pre-shock equilibrium price. If everyone reasons the same way, my decision makes sense because average prices will be about the same as before." A subject in the NTS may start to reason the same way: "Suppose everybody else chooses the same price as before. What should I do? I should choose a price which is much lower than the pre-shock equilibrium price. If everyone reasons the same way, my decision does *not* make sense because then, average prices will be low. In this case, I should choose a high price. If everybody reasons the same way, my decision is still not sensible because then, I should choose a low price." We speculate that the apparently "high degree of consistency in reasoning" may lure some subjects in the NTC to stop searching any further and stick with the choice suggested by the first step of reasoning.[6] Money illusion (along the lines explained in the preceding paragraph) may interact this effect.

Figure 33 shows that there is a clear treatment effect on the percentage of subjects who display correct expectations of the price level for 4 periods after the monetary shock. This clear effect is remarkable since the percentage of correct expectations was considerably lower (-20%) in the substitutes treatment in the last period before the shock. In period 21, this percentage falls dramatically in the complements treatment (-70%age points), but only slightly (-25%age points) in the substitutes treatment.[7]

6. It may also be that other factors explain the higher incidence of equilibrium play in the NTS. For example, the number of dominated strategies is not the same in the two treatments. This implies that the equilibrium can be found by fewer steps of iterated elimination of dominated strategies in the NTS than in the NTC.
7. Accordingly, the loss decomposition shows that efficiency losses due to failure to predict average prices correctly are much smaller in the NTS than in the NTC (see appendix A3 for details).

R22: *The variation of strategic properties affected expectation formation. Expectations are much less sticky in the NTS than in the NTC. High-number illusion or a (psychological) factor inherent in strategic properties may be responsible for this.*

Figure 33: Percentage of subjects holding exactly correct expectations (NTC vs. NTS)

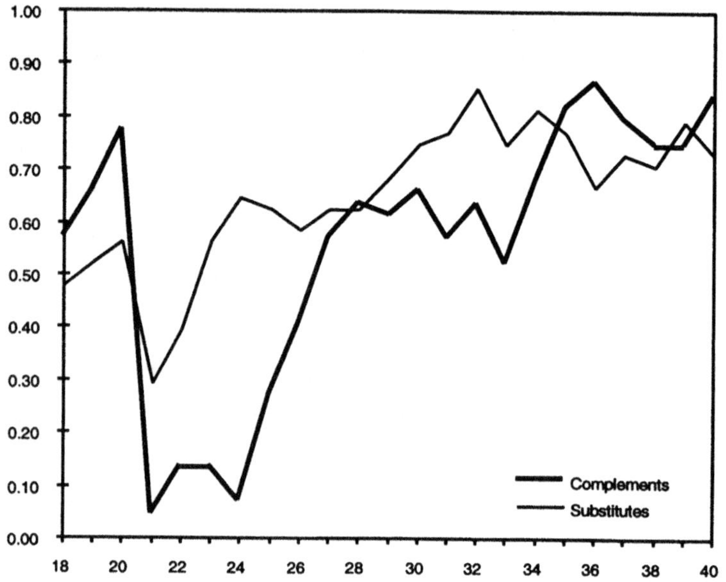

E) Confidence in price expectations

Figure 34 shows that confidence is less shaken by the announcement of the money shock in the NTS than in the NTC. Combined with the data on expectational errors, this supports our earlier interpretation that the subjective indication of the correctness of expectations is a reasonably good indicator to the objective ability to predict average prices. The measure of confidence qualitatively tracks the evolution of the percentage of correct expectations. Hence, subjects express that they find if more difficult to predict aggregate behavior which in fact was the case (see *figure* 33).

Figure 34: Confidence (NTC vs. NTS)

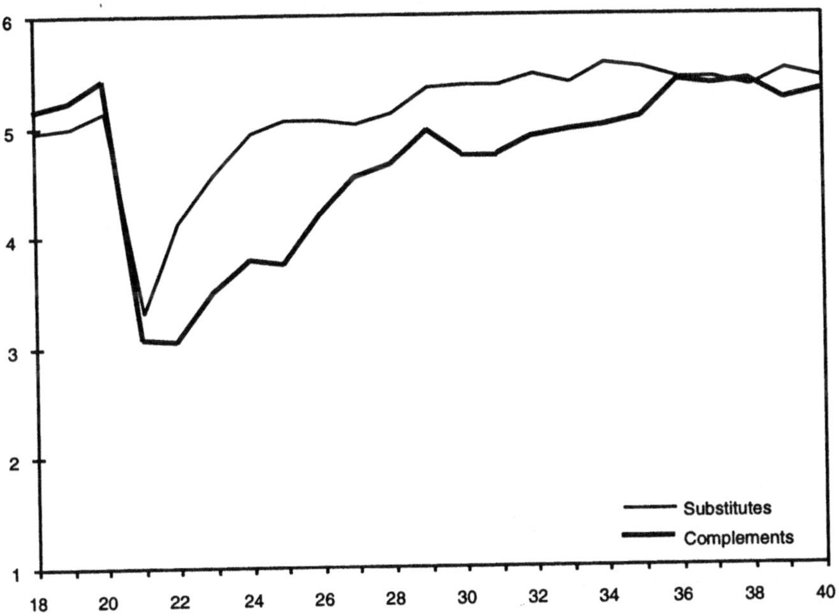

3.3. The effect of strategic properties when the environment is represented in real terms

This section discusses the effects of the strategic properties when the environment is represented in real terms (RTC vs. RTS, see *table* 11). In both cases, the monetary shock can be thought of as a stylized currency reform. The Money illusion hypothesis H_{A1} holds that monetary non-neutrality should be the same in the RTC and RTS. In contrast, the HALTIWANGER / WALDMAN- hypothesis H_{A2} holds that non-neutrality should be larger in the RTC than in the RTS.

The data presented in this section have been generated by the behavior of 84 subjects (40 in the RTC and 44 in the RTC).

A) Nominal rigidity

With respect to nominal rigidity the following result emerges

R23: *The effect of strategic properties on nominal rigidity greatly depends on representation. The effect of strategic properties is large with a nominal representation, but small with a real representation.*

Figure 35 shows the evolution of average nominal prices of the median group when the environment is represented in real terms. In the RTS, the figure shows exactly what one would expect from traditional economic theory: Nominal prices *instantaneously and equiproportionately* adjust to the new nominal equilibrium value. this fact suggests that of the two potential explanations given for the surprisingly fast adjustment of nominal prices in the NTS (see page 132), the one related to money illusion is the less important. However, incorrect anticipation of a mixture of the two elements provoking money illusion may still explain quick adjustment in the NTS.

Figure 35 shows that prices do not instantaneously adjust in the RTC. Yet, price adjustment is very rapid. Do the effects of strategic properties on the speed of price adjustment depend on money illusion? Comparison of *figures* 35 and 26 provides the answer. The latter shows that given the nominal representation nominal prices quickly adjust in the substitutes condition, but exhibit massive nominal inertia in the complements condition. Hence, *figure* 26 shows a massive effect of strategic properties on price adjustment given the nominal representation. In contrast, *figure* 35 shows that strategic properties have a minor effect on nominal inertia given a real representation. Statistical analysis reconfirms the optical impression.[8] This leads us to the conclusion that strategic properties matter more when money illusion is prevalent. Hence, hypothesis H_{A2} appears to be incomplete, and H_{A3} is supported.

8. According to a median test (one tailed, $p < 0.05$) group prices are statistically different in the real representation (RTC vs. RTS) only in the first period after the shock, whereas they are different for 12 post-shock periods according to the same test in the nominal representation (NTC vs. NTS).

Figure 35: Nominal average price of median group (RTC vs. RTS)

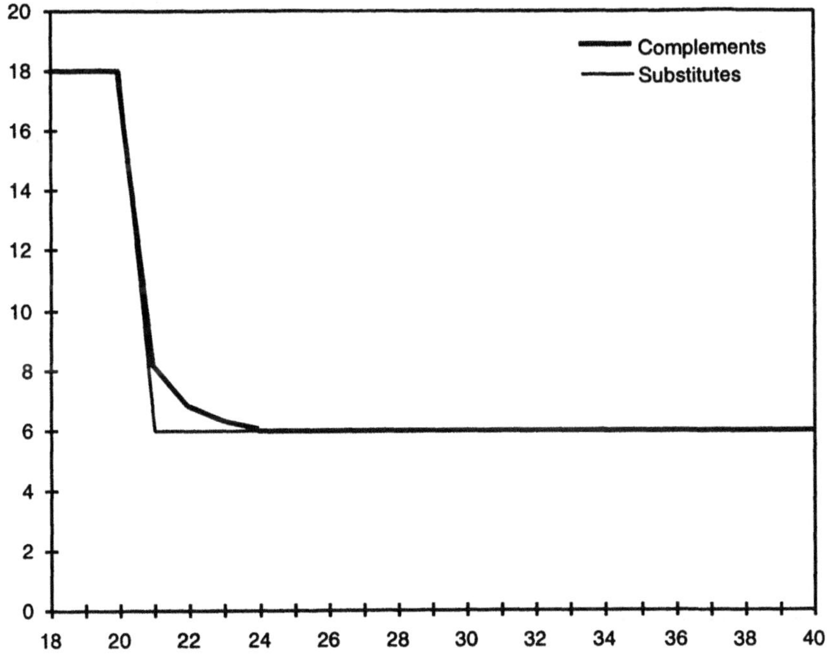

B) Monetary non-neutrality

Figure 36 shows real efficiency losses of the median group. Efficiency losses are larger with strategic complements than with strategic substitutes in either representation. Thus, the hypothesis H_{A2} seems to be confirmed. However, the figure also shows that the impact of strategic properties is smaller with a real than with a nominal representation. In period 21, for example, efficiency losses of the median group are 67% (NTC) versus 10% (NTS) given the nominal representation. In the real representation, the respective efficiency losses are 27% (RTC) vs. 3% (RTS).

R24: *Strategic properties matter more when money illusion is prevalent.*

The comparison of RTC and RTS is consistent with H_{A2} since "natural" inertia (which arises separately and independently of money illusion) is magnified with strategic complements but mitigated with substitutes.

Figure 36: Average efficiency loss of median group (RTC vs. RTS; NTC vs. NTS)

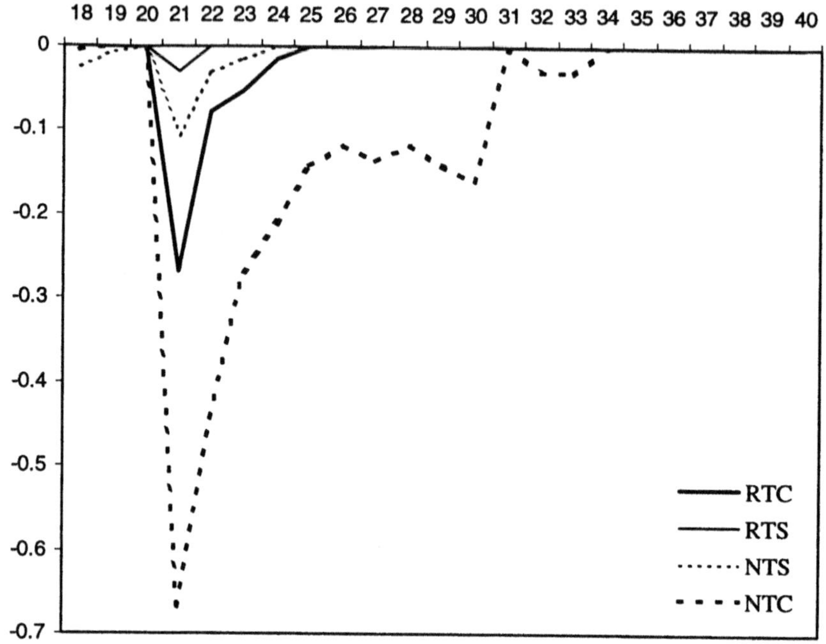

C) Best reply behavior and expectations

Best reply behavior is similar in the RTS and RTC as in all other treatments: Most subjects choose exact best replies given their expectations. We noted in section 3.2. that strategic properties not only affect behavior given expectations (as suggested by H_{A2}) but also affects expectations. We observe that expectations were much more sticky in the NTC than in the NTS (see *figure* 32). In fact, equilibrium expectations were much more prevalent in the post-shock phase in the NTS than in the NTC. This prevalence of equilibrium expectations translates into a prevalence of equilibrium choices in the NTS. *Figure* 37 shows that the percentage of subjects who choose equilibrium prices was considerably higher in the post-shock phase in the NTS than in the NTC.[9] For example, in period 21 70% of subjects chose the equilibrium price in the NTS but only 11% did so in the NTC (see dotted lines). The qualitative effect is similar in the real representation: In period 21, the figures are 82% (RTS) vs. 47% (RTC) (see solid lines). In section 3.2. we hypothesized that the differences in equilibrium

9. This is remarkable because the pre-shock percentage of equilibrium choices was somewhat smaller in the NTS than in the NTC.

expectations and actions may be due to psychological differences which are inherent in strategic properties ("seemingly high degree of consistency" in the CT). We also speculated that this effect may interact with high-number illusion. This conjecture seems to be supported by *figure* 37: The percentages are higher in the ST in both representations, but the differences in these percentages are more pronounced in the nominal representation.

Another interesting feature of *figure* 37 concerns the behavior after period 21. The percentage of equilibrium price choices falls further in the CT in both representations in period 22. This is due to overestimation of rationality of others (as already mentioned on page 89). In the ST we do not observe such a decline in periods 22 and 23 in either representation. Instead, the percentages start to increase again after period 21 in the NTS and RTS.

Figure 37: Percentage of subjects who choose equilibrium prices (RTC vs. RTS)

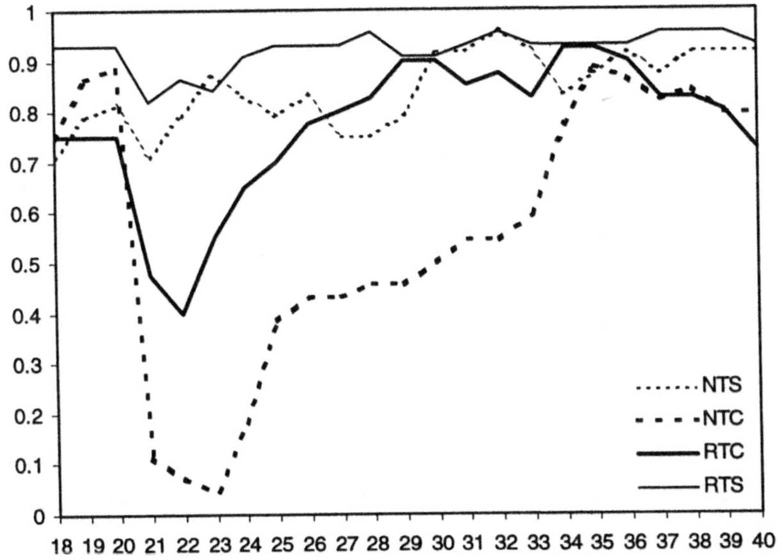

Chapter 4: Summary of results

Chapter 1 discussed the effects of an exogenous, anticipated money shock when the environment is characterized by strategic complements and is represented in nominal terms (NTC). According to the Neutrality hypothesis H_0, an anticipated money shock should be neutral irrespective of the representation or the strategic properties of the economic environment. This hypothesis is clearly rejected. Aggregate-level data show that an anticipated monetary shock causes massive nominal rigidity (see result R1) and monetary non-neutrality (R2): Average nominal prices only adjust to the new equilibrium level after 12 periods, and average efficiency losses are large in the first few periods before the shock (-65% impact effect). The results provide support for the notion of short-run non-neutrality of anticipated monetary shocks, but also for long-run neutrality of money since aggregate behavior eventually converges to the predicted equilibrium (R3). Individual-level data show that subjects behave very rationally given their expectations (R4). Most importantly, price expectations are sticky and are much less correct after the shock (R5). These findings allowed the formulation of a first tentative explanation of the causes of monetary non-neutrality (R6): because expectations were sticky (i.e. subjects expected other subjects to adjust nominal prices only a little) and because subjects chose best replies to their expectations, sticky expectations lead to sticky nominal price choices given strategic complementarity. This downward nominal rigidity translates into real income losses. The loss decomposition supports this explanation. It shows that the efficiency losses mainly result because subjects hold wrong expectations about aggregate price behavior (R7). Consequently, subjects are less confident in their ability to predict the price level. The announcement of the anticipated monetary shock seems to have fundamentally shaken the process of expectation formation (R8).

Chapter 2 investigated whether money illusion is a cause of monetary non-neutrality. The hypothesis H_{A1} was operationalized in section 2.1.

Section 2.2. discussed the effects of money illusion given *strategic complementarity.* Three treatments with a Nominal (NTC), a Semi-real (SRTC), and a Real (RTC) representation have been implemented. The results appear to support the Money illusion hypothesis H_{A1}: nominal rigidity (R9) and efficiency losses (R11) are most pronounced in the nominal representation, and lowest when the environment is represented in real terms. For example, representing the environment in nominal terms magnifies aggregate efficiency losses by a factor of 4, as compared to the real representation. Yet, money is long-run neutral independent of representation (R12). Of the two money illusion creating effects, the high-number illusion is as least as important as the FISHER effect (R10). The findings from the loss decomposition in chapter 1 are corroborated, since the percentage of correct expectations is highest in the real representation. In particular, money illusion causes price expectations to be much more sticky (R14). Hence, money illusion has an impact on the aggregate primarily because it affects expectations and less because subjects are confused themselves ("first-order money illusion"). This conclusion is also suggested by the finding that the percentage of best replies (which is a measure of individual rationality) is not considerably affected by representation (R13). Subjects' confidence in the precision of their price expectation is most shaken in the nominal representation (R15). In all, money illusion (in particular "high-number illusion") has been shown to be a causal factor to explain monetary non-neutrality with strategic complementarity.

Section 2.3. discussed the effects of a variation of representation given *strategic substitutes*. According to the hypothesis H_{A1}, nominal rigidity and the non-neutrality of money should be smaller when the environment is represented in real terms. The results show that this hypothesis is not supported in an overall comparison. For example, aggregate efficiency losses are larger in the semi-real representation (SRTC) than in a nominal representation (NTS). In fact, aggregate behavior exhibits only minor differences with strategic substitutes, and the extent of nominal rigidity (R16) and the non-neutrality of money (R17) are small in both representations.

The comparison shows that "money illusion matters more with strategic complements" (R18) as suggested by the "money illusion *cum* strategic anticipation" hypothesis H_{A3}. The interaction of money illusion and strategic properties was analyzed in more detail in chapter 3.

Chapter 3 investigated whether strategic complementarity is a cause of monetary non-neutrality. Section 3.1. operationalized hypotheses H_{A2} and H_{A3} with respect to aggregate behavior and expectations. This operationalization was based on the results of simulation study.

Section 3.2. discussed the effects of strategic complements vs. strategic substitutes when the environment is represented in nominal terms (NTC vs. NTS). Aggregate-level data show a massive treatment effect: Nominal rigidity (R19) and aggregate real income losses (R20) are massive with strategic complements but very small with strategic substitutes. The (non-interactive) Money illusion hypothesis H_{A1} which holds that monetary non-neutrality should be the same in both treatments is clearly rejected. In contrast, the HALTIWANGER / WALDMAN hypothesis H_{A2} performs relatively well in this comparison. Individual level data suggest that this effect comes about not just because subjects are confused but also because sophisticated subjects aggravate (with complements) or mitigate (with substitutes) the impact of the boundedly rational subjects.

A comparison of observed behavior with simulation results shows that the HALTIWANGER / WALDMAN hypothesis is incomplete. In particular, it is shown that the strategic properties also affected expectations formation. Compared to simulation results, expectations were much more sticky in the NTC than in the NTS. A cognitive explanation for effects of strategic properties on expectations has been provided (R22). Because of this difference, it was not possible to replicate actual aggregate price behavior and real effects with simulations that assume some mixture of "super-rational" (perfect foresight) and "completely naive" (fully adaptive) agents. Hence, real people neither seem to be as rational nor as naive as assumed in theory.

Closer analysis of expectations data shows that expectations were less precise in the complements than in the substitutes treatment. The loss decomposition reconfirms this finding. This fact is also mirrored in the data on subjective confidence, which is lower with strategic complements. Again, individual rationality does not seem to be affected by the variation in the strategic property since best reply-behavior is indistinguishable across treatments.

Section 3.3. discussed the effects of strategic properties when the environment is represented in real terms. Strategic properties do have the effect predicted by H_{A2}. However, nominal rigidity (R23) and monetary non-neutrality are much more pronounced in the nominal than in the real representation. Thus, hypothesis H_{A2} is incomplete since it neglects the effects of representation. The "money illusion *cum* strategic anticipation" hypothesis H_{A3} organizes the data very well: the strategic properties matter more when money illusion can be supposed to be prevalent (R24).

Overview. *Table* 15 shows aggregate efficiency losses in the 10 periods after the monetary shock (normalizing the losses in the NTC treatment to 100). As can be seen, efficiency losses are largest in the NTC (H_0 is clearly rejected: $100 \neq 0$). Systematic variation of the representation holding the property of complementarity constant reduces real effects of the monetary shock: Each of the two components of the veil of money halves the real effects (H_{A1} confirmed: 100 > 55 > 27). However, the money illusion hypothesis is incomplete since real income losses are larger in the SRTC than in the NTS (55 > 40). Monetary non-neutrality is larger with strategic complements than with substitutes (H_{A2} confirmed: 100 > 40; 27 > 13). However, these strategic properties matter more when money illusion can be supposed to be important (H_{A2} rejected: 100 - 40 > 27 - 13). In addition, money illusion matters more with strategic complements (H_{A1} rejected: $100 \neq 40$; $27 \neq 13$). Remarkably, the variation in both dimensions shows that the non-neutrality almost disappears when the environment is represented in real terms and is characterized by strategic substitutes. Hence, we found an experimental environment (albeit a very "unnatural" one) in which the Neutrality hypothesis H_0 predicts behavior well. Starting from this environment, neither strategic complements nor money illusion alone are able to produce considerable short-run non-neutrality. The combination of these two very "natural" aspects of macroeconomics, however, causes pronounced non-neutrality. This massive effect is unexpected by traditional macroeconomics since it is observed in an environment which is friction-free and after a fully anticipated monetary shock. In all, the hypothesis H_{A3} that "strategic properties matter more when money illusion is prevalent" and "money illusion matters more with strategic complementarity" is supported.

Table 15: Comparison of aggregate efficiency losses (normalized)

Relation of aggregate efficiency losses (period 21 - 30)	Nominal representation	Semi-real representation	Real representation
Strategic complements	100	55	27
Strategic Substitutes	40	-	13

Part IV

Discussion of Results

The results of this study show that money illusion has prematurely been dismissed as an explanation of nominal rigidity and monetary non-neutrality. This conclusion is based on theoretical considerations and empirical findings. At the theoretical level, we argued that in order to rule out the relevance of money illusion it is not sufficient that all individuals are illusion-free. Instead, one has to assume that the absence of money illusion is common knowledge. Yet, this assumption is very unlikely to be met in practice. On the empirical level, it has been shown that money illusion causes massive monetary non-neutrality if strategic complementarity prevails (for a summary of results, see Part III, chapter 4).

Chapter 1 discusses the empirical relevance of the experimental results. In particular, it is discussed which features of the experimental design and its procedures may have biased the results in one way or another. The issue of how the environment in the experimental laboratory differs from actual business life is addressed. Suggestions for further experimental research are provided.

Chapter 2 discusses implications for economic theory and policy. It is argued that the rationality paradigm should be used as a benchmark case to investigate which institutions are robust with respect to individual rationality. The conclusions for economic policy necessarily remain highly speculative. A general suggestion is that automatic stabilizers may decrease the degree of strategic complementarity in the economy. With respect to the conduct of monetary policy, the following, rather general conclusions emerge: monetary contractions should be avoided, monetary authorities should try to coordinate expectations, and the Euro should be introduced with care.

Chapter 1: Empirical relevance of results

"Recipe for Pseudo-Anomalies

Ingredients

1. A common and useful rule of thumb ...
2. A laboratory environment that evokes the rule of thumb when it is inappropriate
3. Subjects inexperienced in the laboratory environment.
4. Standard laboratory procedures.
5. Standard statistical techniques and standard rhetoric.

Directions

Put subjects in the laboratory environment. Run them through the standard procedures, carefully excluding any opportunities for them to learn from their experience or from imitating more successful subjects."

DAN FRIEDMAN (1998: 941)

A prominent experimentalist recently criticized that experiments are sometimes abused to demonstrate "pseudo-anomalies" (see quotation above). The author pointed out that observed behavior may very much depend on the specifics of an experimental design and its procedures. In particular, FRIEDMAN (1998: 941) states that "Every choice anomaly can be greatly diminished or entirely eliminated in appropriately structured learning environments." We completely agree with this assertion. In fact, this study provides an example for this statement. We have identified conditions (strategic substitutes) under which the "choice anomaly" of money illusion does *not* considerably affect aggregate behavior. Controlled variations of the decision environment allowed the identification of other conditions (i.e. strategic complements) under which money illusion does have massive effects on aggregate real income. The experimental method therefore allowed the isolation of money illusion and strategic complementarity as causes of monetary non-neutrality. We suggested that the experimental results are highly relevant because strategic complementarity and the use of money are very natural features in actual macroeconomies. However, one may claim that the results are in some sense biased because all treatment conditions create an environment which is favorable to the finding of monetary neutrality, i.e. *against* the finding that money illusion matters. Such biases arise if the design deviates in important respects from naturally occurring economies. Since we consider this criticism to be important, this chapter takes another look at the specifics of the design and the procedures with respect to external validity of the findings (see also Part II, chapter 2).

The discussion in this chapter proceeds as follows: Section A) explains which aspects of the design bias the results *in favor* of the finding of monetary neutrality, and explains why these biases have been introduced. However, some aspects of the design may be suspected to create biases *against* the finding of monetary neutrality. Section B) discusses these aspects. Section C) provides suggestions for further experimental investigations in this field.

A) Biases in favor of monetary neutrality

A general critique of experiments on "anomalies" is that anomalous behavior will diminish or disappear when experiments are designed to give subjects stronger incentives, greater expertise, better opportunities to learn, unchanging circumstances, and a simple context (see CONSLIK, 1996). Thus, favorable conditions may create an environment in which people come to act "as if" they were smarter than they are. However, the learning logic does not *per se* provide insights into the empirical ("external") relevance of experimental findings since the logic cuts both ways. As will be explained below, conditions were very favorable for learning in our experimental design, but are they in the "real world"? As EINHORN and HOGARTH (1978) have emphasized, many situations do not provide feedback in a way that facilitates learning. How many monetary contractions (or expansions) do we witness in our lives? How often do firms revise their prices (see section I/4.2 for evidence)? Does any firm know its profit function and the profit functions of all their competitors? How frequently do firms get unambiguous feedback (in terms of opportunity losses) in response to pricing decisions?

Why does the design deliberately implement unrealistic conditions facilitating learning if these conditions create biases in favor of monetary neutrality (i.e. biases against the finding that money illusion matters)? The reason is an *a fortiori* argument: if we find that money illusion matters despite these biases, it is highly probable that money illusion will also matter in environments which are less favorable to monetary neutrality. Put differently, these biases have been introduced to be "on the safe side" with the statement that money illusion matters.

The following sources of biases are now discussed: Participants in the experiment were not representative of the population at large, and they were trained to overcome money illusion. The decision situation in the laboratory is much more simple than the decision situation of a firm in business life. Parameter choices which make the situation cognitively simple concern the size of the shock and of the experimental economy. The decision environment in the laboratory is (exogenously) stable, rules out (exogenous) uncertainty and provides good learning opportunities. Subjects' information set was well-structured, informational feedback was immediate and reliable.

Subject pool bias. Undergraduate students from all faculties (except economics majors) were recruited as experimental subjects. This selection of experimental subjects may have biased the results in favor of monetary neutrality because students are presumably more skilled and certainly more apt to perform divisions, and thus less prone to money illusion than the average person in the population. However, we can make an *a fortiori* argument: if money is non-neutral with these above-average skilled subjects, it should be even more so with people of average cognitive skill.

Exercises. All subjects had to solve several control questions before the experiment started. This included exercises to read the payoff tables and (in the nominal representation) the correct deflationing of nominal payoffs. Moreover, before the experiment started, all subjects were publicly informed that everyone was able to solve the exercises correctly. This procedure creates the precondition for the common knowledge of the absence of money illusion. Hence, this procedure biases results against the finding that money illusion matters. The usual *a fortiori* argument applies.

Size of shock. The experimental design implemented an exogenous (negative) monetary shock which is anticipated and common information. The negative shock was implemented to be able to discriminate the money illusion hypothesis from other explanations of sticky price adjustment (see page 75 for explanations). The parametrization was chosen to make the decision situation as cognitively simple as possible for subjects. For example, only discrete prices from a given interval could be chosen and an agent's own price was excluded from the average price that is relevant for this agent. The huge size of the monetary shock is a consequence of parametrization choices which were aimed at rendering the decision environment as simple as possible. Since the implemented equilibrium nominal prices are $\bar{P}_0^* = 18$ and $\bar{P}_1^* = 6$ respectively, this implies a reduction in the quantity of money of $M_0/M_1 = 3/1$ (see appendix A3). To be able to (statistically) observe movements of prices, subjects were given the choice among a relatively wide range of prices $P_i \in \{1, 2, ..., 30\}$. If the parametrization were implemented, for example, in the range

$P_i \in \{1001, 1002, ..., 1030\}$, the same tables with exactly the same entries could have been used, but now the implied monetary shock would have been only about 1% (= 12 / 1018).

Size of experimental economy. Obvious practical reasons limit the number of potential participants in an experiment. In the present design, the size of the experimental economy was very small ($n = 4$) but the number of observations was quite large (in all, 216 subjects, i.e. 54 independent group observations). Taking into account obvious financial and time constraints, a trade-off arises: for a given number of experimental subjects (i.e. expenses), a small (large) group size generates many (few) independent observations. We preferred to get many independent observations because this allowed us to test statistically for differences between treatments. However, it is not clear whether the relatively small group size biases the results in favor of monetary neutrality or non-neutrality. There are two countervailing effects: on the one hand, the impact of a single individual on other players' average price is proportionately lower with larger groups. Thus, an agent prone to money illusion has a smaller direct impact on the aggregate in larger groups. On the other hand, assume that a given proportion of the population is prone to money illusion. In this case, the probability increases with group size that a money illusion prone subject is in the particular group. However, the issue of group sizes may lose its weight when considering that all treatments were run with the *same* number of participants (per group) and aggregate behavior displayed large differences among treatments.

Matrix representation and dominated strategies. The only (formal) difference between the complements and the substitutes treatment is that we change the sign of the slope of the reaction function (see appendix A3). However, due to the limited support of the matrix representation, one could argue that the comparison of strategic complements and strategic substitutes does not constitute a true *ceteris paribus* variation. The reason is that the number of dominated strategies is not the same in both post-shock tables. This property may additionally help subjects to find the equilibrium and hence bias results in favor of monetary neutrality in the substitutes treatment. It may be interesting to investigate how the number of dominated strategies affects nominal inertia under the respective strategic properties.[1]

Learning and information feedback. Learning to behave optimally is facilitated when economic agents are provided with sufficient information about the environment and with well-structured, immediate informational feedback about the consequences of their decisions. The design is biased in favor of monetary neutrality because the information conditions were very unrealistic in the sense that they were much better than in the "real world". For example, subjects knew their own and their opponents' payoff functions. In business life, firms usually do not know their competitors' payoffs (not to speak of the payoff *functions*). In business life, firms may not even know their own payoff functions. In addition, the experimental design provides decision makers with immediate informational feedback about their actions. At the end of each decision period, each subject is told which prices the other firms chose and what their own real payoff was. The decision situation is also very simple in the laboratory because firms do not have to solve intertemporal optimization problems (like saving or investing).

Stationary replication. In the present design experimental subjects had to make 20 decisions in the pre-shock table, and again 20 decisions in the post-shock table. The design created a stable environment without any exogenous uncertainty. In the design there are, for example, no random cost shocks or otherwise unexpectedly changing conditions. Repeating choices under the same conditions (so-called stationary replication) facilitates learning and, thus, convergence to the Nash equilibrium. In fact, aggregate behavior converged to the equilibrium in all treatments after some periods. Of course, we could have implemented a design in which agents make only one decision in the pre-shock, and one decision in the post-shock table. However, such a procedure would not allow to observe whether and how fast prices converge to the equilibrium.

Strategic substitutes. The anticipated money shock was approximately short-run neutral with strategic substitutes, irrespective of representation. However, strategic substitutes are very untypical for a pricing game and for most macroeconomic relations. Hence, behavior in these treatments should be viewed as the "unnatural" case which mainly serves as a control treatment to isolate the effects of strategic complements.

1. However, a first pretest where the number of dominated strategies was held constant across complements vs. substitutes treatments produced results which are qualitatively very similar to the findings of Part III, chapter 3. Hence, the results are surprisingly robust in this respect.

Pareto-efficient equilibrium. In the present design subjects cannot gain from cooperation, i.e. by deviating from the equilibrium. This is a very unusual feature of a pricing game. In pricing games the equilibrium will typically be inefficient (because high prices depress aggregate demand), and a collective price reduction may increase welfare. We have good reasons to implement an inefficient equilibrium though: we know from many experimental studies that subjects try to reap the gains from cooperation in games with inefficient Nash equilibria (see e.g. LEDYARD, 1995). If we implemented an inefficient equilibrium, there would have been a second force (in addition to money illusion) inhibiting adjustment to equilibrium. In this case, it would not have been possible to properly isolate the effects of money illusion.

Short run vs. long run. In some sense, the results of this study show that money illusion does not matter in the long run. We arrived at this conclusion because aggregate behavior converged to the money neutral equilibrium in each representation after some periods. How can we than judge whether the effects of money illusion are "long lasting" or not? The answer depends on the parallelism of the informational feedback in the experiment and the "real world". In the experiment, subjects received immediate, precise and rich informational feedback. In business life, it is difficult for firms to learn what the consequences of a pricing decision are. In some markets, firms may only learn at the end of the year whether real profits were higher or lower in that year. Sometimes it seems to be almost impossible for a firm to know whether or not profits were maximal and even more so whether deviations from the maximum were caused by a pricing decision or by some other factor.

B) Biases against monetary neutrality?

In section A) we argued that conditions in the laboratory were more favorable for quick convergence to the rational prediction than in actual business life. The experimental results support this claim since aggregate behavior converged in all treatments, and in some treatments behavior even converged instantaneously. A critic may agree with this, but insist that conditions could have been implemented which are even more favorable for the finding of monetary neutrality, i.e. which create even stronger biases against the finding that money illusion matters. Such a critic may suspect that money illusion would not cause monetary non-neutrality is the case even when strategic complements prevail. An alternative criticism is the following: money illusion would not matter in a more realistic (i.e. more complex and hostile to learning) environment because people would engage in some sort of collective action (e.g. communicate). Despite complexity, it is sometimes argued, this would allow instantaneous coordination on the equilibrium.

Learning and repetition. The present design implemented one single monetary contraction. M changes from M_0 to M_1 at the end of period 20 and remains at that value throughout the post-shock phase ($t = 21$-40). One could argue that the subjects would have learned to avoid money illusion when experiencing several shocks in a row. However, it is impractical for experimental reasons to implement several shocks in a row because this just takes too long if the stationary replication technique is applied. For example, it took approximately two hours to be sure that prices converge in the pre-shock phase and to implement a single shock. If an experiment takes more than three hours, the danger arises that subjects become tired, unconcentrated or even bored. A more reasonable way to investigate the effects of repeated shocks would be to use experienced subjects. That is, subjects which already participated in one of the sessions could be called again at some later date to participate in a similar experiment (with a different parametrization).

Cognitive complexity. A critic may claim that the experiment was cognitively very complex in the nominal representation. After all, subjects are given two 30 x 30 tables which makes 1800 payoffs to consider in total. No student, the critic may argue, is able to perform 1800 divisions to uncover the real payoffs to determine the equilibrium choice within the given time frame. This criticism is based on a misunderstanding. It is not necessary to perform the divisions from a game theoretic perspective because the best reply structure of the game, of course, remains unaffected by displaying nominal payoffs. Therefore, the cognitive complexity is the same in the real and the nominal representation to find the best replies. Subjects just have to look for the highest (nominal or real) payoff for each average price. In fact, an overwhelming majority of subjects chose best replies in all treatments. Moreover, 1800 nominal payoffs were shown in both the NTS and the NTC. Despite this fact, aggregate behavior almost perfectly adjusted to the monetary shock in the NTS.

Degree of strategic complementarity. The main purpose of this study was to test whether strategic complementarity in conjunction with money illusion causes money to be non-neutral. An important parametrization choice concerns the degree of strategic complementarity that is implemented. In fact, we chose the maximum possible degree of strategic complementarity without obtaining multiple equilibria. According to the HALTIWANGER / WALDMAN hypothesis the effects of the monetary shock should be smaller when the degree of strategic complementarity is reduced.[2] This hypothesis is easily testable in the present design, and it may be interesting to investigate an intermediate case with weak strategic complementarity (i.e. a slope of the reaction function closer to zero). However, implementing a reaction function with slope exactly equal to zero may not provide a very interesting case. This would correspond to a situation without strategic interaction, i.e. an individual decision-making experiment.

Asymmetric agents. The experimental design implemented two types (x- and y-types) of agents. These agents differed exclusively with respect to their equilibrium choices. One type had to choose relatively low prices, the other type relatively high prices in equilibrium. This property of the design can be considered to be relatively realistic, since heterogeneity is typical in business life (e.g. price competition between producers of branded products). If homogeneous agents were implemented the agents would have been provided with an additional way to find the equilibrium which is not available in business life. (The equilibrium can in this case be found graphically, as the intersection of the best reply with the 45-degree line).

Simultaneous choice. In the design, firms had to choose prices simultaneously. It is unclear whether this feature introduces a bias in favor or against monetary neutrality. If price choices were made sequentially within each decision period, aggregate behavior would probably strongly be path dependent (the path would depend on the choice of the first firm). It may be interesting to investigate how sequential price choices affect the aggregate.

High stakes. One could argue that exposing subjects to higher incentives would considerably attenuate aggregate effects of money illusion. This, of course, is an empirical question, and it is easily testable. However, it should be noted that many situations which may involve money illusion do not provide high stakes (e.g. many consumer choice problems). In general, some experiments find that anomalous behavior is reduced with high stakes, but many do not.[3]

---

2. See proposition 2 in HALTIWANGER and WALDMAN (1989).

Communication. In the experiment, subjects were not allowed to communicate with each other. Yet, one may suspect that prohibiting communication introduces a bias against monetary neutrality because communication could facilitate learning, promote coordination and, hence, mitigate nominal rigidity. Luckily, this hypothesis is easily testable. One could just run the same experiment but let subjects talk to each other before each decision.[4] However, it is not clear whether allowing communication would add "realism" to the experiment. How are millions of people in an actual macroeconomy supposed to talk effectively to each other? Of course, people in some sense communicate about the effects of monetary policy in newspapers and TV. Everyday experience suggests, however, that this does not necessarily lead to perfectly coordinated behavior or even consensus about macroeconomic issues.

C) Extensions and suggestions for further research

In addition to discussing empirical relevance of experimental findings, the two preceding sections also illustrate very attractive features of the experimental method. First, experimental results can be *replicated* by any critic.[5] This guarantees that idiosyncratic experimental handling (so-called experimenter demand effects) which may have affected results can be detected. Second, the *robustness* of findings can be investigated since virtually all criticisms mentioned above are easily amenable to experimental testing. It can then be tested (experimentally) whether these factors are indeed important. Critics are encouraged to run their own experiments and to *provide evidence* in favor of their objection. In fact, many open questions remain, and the investigation of the causes and consequences of money illusion deserves further research effort. Below, suggestions are made to investigate the following questions: Is it possible that money illusion causes long-run non-neutrality? How important are other potential causes of nominal rigidity as compared to money illusion? Which cognitive

3. A prominent example is structured bargaining in the so-called Ultimatum game in which the game theoretic prediction systematically failed to predict behavior (GÜTH, SCHMITTBERGER and SCHWARZE, 1982). To investigate whether increasing incentives ("high stakes") would systematically affect bargaining outcomes, CAMERON (1995) conducted experimental sessions in Indonesia where subjects could earn several months' income within a few minutes. In this case, high stakes did not affect "anomalous" behavior. However, significant differences were observed between trials where subjects were asked hypothetical questions and paid experimental sessions. See also KACHELMEIER and SHEHATA (1992) or FEHR and TOUGAREVA (1996).
4. One could speculate that a person with superior mental capacities could convince everybody else to do as he or she suggests. Indeed, this is to be expected in the present design since the number of subjects is small and the equilibrium is Pareto-efficient.
5. Instructions are included in appendix A1, the tables are to be found in appendix A2. In addition, the program to run the experiment is available from the author upon request.

features induce money illusion? These questions can be addressed with slight adaptations of the present design.

Multiple equilibria. The present design implemented a unique (and pareto-efficient) equilibrium because we focus on the causes of short-run non-neutrality. It would be interesting to investigate whether nominal shocks could also lead to *long-run non-neutrality* if, for example, equilibrium selection were affected by money illusion. Money illusion could hence serve as a "coordination device". If the equilibria are pareto-rankable, and the pareto-dominated equilibrium happens to be at high nominal (but low real) payoffs, one may test whether high-number illusion leads to long-run (i.e. stable) real income losses. The design can easily be adapted to investigate this question, and it is planned to conduct corresponding experiments soon.

Near-rationality. The present design did not favor near-rational behavior in the sense that small deviations from individual rationality (i.e. choosing an exact best reply) lead to small (second order) losses. However, the design could easily be adapted to investigate this alternative explanation of nominal inertia.

Exogenous nominal frictions, staggering. The results of this study show that money illusion has been prematurely dismissed as an explanation of monetary non-neutrality. However, the study does *not* contest the usefulness of alternative explanations of nominal rigidity. The present design can easily be adapted to test the hypotheses that staggering (e.g. TAYLOR, 1979) or menu cost are causes of nominal inertia (e.g. MANKIW, 1985). To test the staggering hypothesis, one could run exactly the same experiment (with strategic complements) but allow only a fraction of subjects to change their price in a given period. Depending on the fraction, subjects would only be allowed to change their prices every few periods. To test the menu cost hypothesis, one could run the same experiment, but now simply charge a small fee for each price change (see WILSON, 1998 for an alternative approach).

Asymmetric effects of money. Negative shocks were implemented because it was suspected that subjects may suffer from high-number illusion. Since nominal payoffs are higher with higher average prices, this effect works against monetary neutrality when the environment is represented in nominal terms and strategic complementarity prevails. If this hypothesis is correct, nominal rigidity and real effects should be much less pronounced after a *positive* shock. Given strategic complements, the high-number illusion would induce the choice of high nominal prices (which are close to the equilibrium). Positive money shocks can easily be implemented in this design. However, they would also lead to efficiency losses because the equilibrium is efficient. Thus, the hypothesis that a positive nominal shock yields efficiency gains cannot be tested with the present parametrization. Further research should be devoted to the question whether money illusion may explain why monetary shocks may have asymmetric real effects.

Anticipation of shocks. Anticipated shocks were implemented because this study was aimed at testing the Neutrality hypothesis suggested by the New Classics. However, it would have been easy to implement unanticipated shocks,[6] but not very telling. Considering the evidence discussed in Part III, these would almost certainly be non-neutral in either condition.

Isolating aspects of the "veil of money". Two hypotheses about why money illusion may affect the aggregate have been tested: the high-number illusion (deflationing effect) and the FISHER effect (smallest nominal accounting unit effect). In the RT, none of these elements is active, in the SRT only the FISHER effect is operative, and in the nominal representation both effects are operative. Surprisingly, the high-number illusion turned out to be the more important aspect of money illusion. However, in the present design the high-number illusion interacts with nominal anchoring (see also Part III, page 102). It would also have been interesting to test the effect of high-number illusion in isolation. To this end, one would have to run the "Real" treatment (RT) with nominal payoffs shown in the payoff matrix. A comparison of this treatment to the original RT could provide additional insights on the interaction of high-number illusion and nominal anchoring.

First- and higher-order money illusion. The effects of first- and higher-order illusion can only be inferred indirectly in this design. A possibility to *directly* discriminate between first- and higher-order money illusion is the following: each subject is given the same instructions and payoff tables as in the present study, but each subject is told that he or she plays against three computers. Each subject is told that these computers are programmed to choose the best replies given the action of the subject. Thus, the computer simulates super-rational opponents. These are not only rational enough to

6. For example, one could run exactly the same the experiment without announcing that something has changed and without distributing new payoff tables at the end of period $t = 20$.

know what the equilibrium is, but also clever enough to know to what extent the opponent is boundedly rational, i.e. it is known to the human players that these computers simulate players with perfect foresight. Since the equilibrium is efficient there is no reason whatsoever for a rational human subject not to choose the equilibrium price. Some sessions could be run with a real, some sessions with a nominal representation. The differences in monetary non-neutrality would then be caused by "first-order money illusion".

Chapter 2 : Implications for economic theory and policy

"You can fool all the people some of the time, and some of the people all the time, but you cannot fool all the people all the time."

ABRAHAM LINCOLN (1858)[7]

This study has shown that money illusion and strategic complementarity are causes of nominal rigidity and monetary non-neutrality. This finding brings up a difficult issue: if the rational-agent paradigm fails to predict behavior in this case, should we give up rational-agent theory? Section 2.1. explains why the traditional answer to this question has been negative. It is argued that a more differentiated answer could be productive. Section 2.2. specifies what providing a "differentiated" attitude to rationality issues implies for macroeconomic theory. Section 2.3. draws some tentative conclusions for economic policy.

2.1. Money illusion and the rationality paradigm

"Should the facts be allowed to destroy a good story?"

MICHAEL LOVELL (1986: 120)

The rational-agent paradigm in fact is "a good story" since it proved to be a very useful and successful paradigm. It is useful by virtue of its transparency and consistency because it helps us to think in a structured way about complex economic issues.[8] It is successful because it allows the prediction of aggregate behavior in many instances and therefore provides many important insights in the social sciences. The rational-agent paradigm has been considered to be empirically successful in predicting aggregate behavior despite the fact that some of its (individual) rationality assumptions obviously are descriptively wrong. It is by now widely accepted in the economics profession that there are limits on human cognition, that people make mis-

7. Attributed words in speech of Sept. 8, 1858. Quoted after J.M. COHEN and M.J. COHEN (1960, eds.): Penguin Dictionary of Quotations. Middlesex 1960: 235.
8. Note that experimental economics has, at least in part, become a powerful tool of economics because of the rational-agent paradigm. Theory testing without a theory which provides clear predictions is just not possible.

takes, and there are heuristics and biases.[9] Despite this acceptance, economists have been traditionally reluctant to include bounded rationality in economic analysis. The reasons for this reluctance are explained in the following paragraphs.

As if-approach. A popular argument in defense of the rationality paradigm is the following: the question is not whether people are in fact perfectly rational, but, rather, whether they act approximately *as if* perfectly rational (FRIEDMAN, 1953). As STIGLER[10] suggested, one justification for this "as if" approach is that the models assuming fully rational agents are good predictors. In other words, economic theories can afford to be descriptively wrong at the individual or the psychological level, as long as they yield good predictions at the aggregate (market or macroeconomic) level. This argument rests on the assumption that there will be only few agents who behave in a boundedly rational manner, and/or that their behavior will not importantly affect aggregate outcomes. In fact, GODE and SUNDER (1993) provide the example of double auction markets in which "zero-intelligence traders" (computers which bid randomly) achieve near perfect market efficiency. This study illustrates that the trading institution of double-auction market generates aggregate results which can be rationalized by assuming that all agents are perfectly rational, despite the fact that they have "zero intelligence". That is, double-auction markets appear to be extremely robust with respect to individually irrational behavior. However, the way by which individually irrational behavior translates to aggregate outcomes does not follow simple patterns. For example, the theoretical argument of this study made clear how extremely restrictive it is to rule out that money illusion affects aggregate behavior (see Part I, chapter 4). To paraphrase LINCOLN (see quotation on page 159), it is not sufficient to rule out that "some people can be fooled all the time", it is not even sufficient to rule out that there are "fools" at all. Rather, one has to assume that absence of "fools" is common knowledge. The experimental results of this study have shown that representation effects which induce money illusion at the individual level indeed translate to substantial effects at the aggregate level in some environments, but may have only minor impact in others. Put differently, strategic complementarity is an "institution" which is

9. For example, GARY BECKER says in his Nobel lecture (1993: 386): "Actions are constrained by income, time, imperfect memory and calculating capacities, and other limited resources".

10. GEORGE STIGLER (1966: 6): "When we assume that consumers, acting with mathematical consistency, maximize utility, therefore, it is not proper to complain that men are much more complicated and diverse than that. So they are, but, if this assumption yields a theory of behavior which agrees tolerably well with the facts, it must be used until a better theory comes along."

extremely susceptible to individually irrational behavior. Hence, the traditional *as if*-argument does not seem to hold concerning money illusion.

There are two arguments in defence of the as if-approach: one is that people learn to overcome money illusion through learning or in high-stakes situations. Some of these issues have been touched upon in the discussion of the empirical relevance of experimental results (see chapter 1). The other commonplace defence is that competitive pressure should eliminate money illusion over time, as e.g. argued by ALCHIAN (1950).[11]

Discipline in theorizing. Some economists argue that relaxing the assumption of full rationality takes away the discipline in theorizing. Without the discipline of optimizing models, they argue, economic theories would degenerate into a collection of *ad hoc* hypotheses which "explain" everything but which lack overall cohesion and scientific refutability. This position is correct insofar as there is usually only one well-defined way to be rational but many ways to be irrational. Allowing the relaxation of the assumption of full rationality in *arbitrary* ways would indeed be like opening Pandora's box. However, experimental economics can act as a discipline device. It provides the tools for a *controlled* and empirically disciplined relaxation of the rationality assumption since (as in this study) particular alternative hypotheses can be tested.

Bounded rationality vs. information costs. Because of the arguments discussed above, theories relaxing the assumption of full rationality are not (yet) very popular among (macro-)economists. In contrast, many researchers have investigated the consequences of information costs. This approach seems to be widely accepted by the economics profession. Yet, closer examination reveals that the two concepts may in fact be very similar.[12] A famous example of the similarity of deliberation cost and infor-

11. It is an apparent paradox that competition may indeed *protect* irrational individuals from being "money pumped" (example by MATTHEW RABIN in a personal communication): suppose an agent i owns an asset (e.g. a piece of gold) worth x, but is completely ignorant of its value. If there is only one bidder who knows x, i will sell the asset for nothing. If several bidders compete, they are both willing to pay x which will be the market price (in many auctions, and if they fail to collude). Hence, agent i is protected from being money pumped by competition as long as he is rational enough to sell to the highest bidder.
12. CONSLIK (1996: 690) points to the similarities of deliberation cost and information cost by formulating: "When I walked into a post while watching a bird, my family called it a dumb move. Among economists, however, I could have claimed that, given the spatial distribution of lamp posts, the expected utility of bird watching exceeded the expected disutility of a collision. *Ex ante*, the post was probably not there, and it is entirely rational to collide with an *ex post* post. This example illustrates the confounding of rationality issues with information issues. Am I dumb to walk into a post or merely a rational victim of imperfect information?"

mation cost is the "Gang of Four" explanation of cooperation in finitely repeated prisoner's dilemma situations (KREPS *et al.* 1982): the authors suggest a possible rescue of standard (rational choice) theory by putting the bound on information instead. They assume that, although both players in fact are fully rational, one player thinks the other might be boundedly rational. Thus, the absence of common knowledge of rationality is enough to induce cooperation with the usual rational assumptions. Similarly, it is sufficient to obtain the results of HALTIWANGER and WALDMAN (1985, 1989) if the assumption of common knowledge of rationality is violated. The present study shows that things are more complicated than that: some agents in fact seem to be prone to first-order money illusion and many seem to be illusion-free but strategically react to others' money illusion. However, most agents are unable to correctly predict the extent to which others are confused or believe that others suffer from money illusion.

We should not dismiss rational-agent theory altogether because it is a theoretically extremely useful paradigm. However, we should not blindly believe that the rational-agent paradigm will always[13] provide reasonably accurate predictions of aggregate behavior. Rather, the rational-agent paradigm should be used to generate clear-cut Null hypotheses and one should use powerful empirical (e.g. experimental) methods to investigate which "institutional arrangements" are robust with respect to (specific types of) individual irrationality.

13. For example, BECKER (1976: 14) states that "...all human behavior can be viewed as involving participants who maximize their utility from a stable set of preferences and accumulate an optimal amount of information."

2.2. Implications for macroeconomic theory

"... slight departures from unbounded rationality have important macroeconomic implications, and models which take such departures into account are better equipped to explain several aspects of observed reality."

RAJIV SEHTI and REINER FRANKE (1995: 583)

This section very briefly relates the results of this study to New Classical and New Keynesian theories of nominal rigidity. Some brief comments on currency reforms are made. As should be clear by now, it is our conviction that including behavioral and cognitive aspects into macroeconomics is fruitful. However, we will abstain from belaboring this rather obvious point (see also quotation above).

New Classics vs. New Keynesians. As discussed in Part I of the book, there are conflicting intuitions about the effects of money on real economic activity, and different theorists have drawn diametrically opposed conclusions from their models. Yet, it seems to be the case that this issue cannot be resolved on empirical grounds without using experimental methods. This is particularly true with regard to the problem of isolating causal relations among aggregates which are difficult to measure. This study has contributed clear empirical evidence to this debate. It shows that the New Classical view provides reasonably precise predictions under some (observable) circumstances, but is completely misleading (in the short run) under others.[14] Unfortunately, the conditions under which the (super-)rational model predicts behavior well are very unlikely to be met in practice. Strategic complementarity is a key characteristic of naturally occurring economies[15] and almost all business transactions are "shrouded in the veil of money". On the other hand, the New Keynesians are right in their prediction that

14. Or, to use the words of HALTIWANGER and WALDMAN (1985: 328): "... for the analysis of situations that exhibit [strategic substitutes], there are relatively strong justifications for assuming rational expectations. However, for the analysis of situations that exhibit [strategic complements], the practice of assuming rational expectations would seem less defensible."

15. As first argued by COOPER and JOHN (1988) and later by OH and WALDMAN (1994) strategic complementarity captures the essence of spillover effects and has been modeled in many different ways. Examples include monopolistic price competition (KIYOTAKI, 1988), search activities in labor markets (HOWITT, 1985; SUMMERS, 1988; DRAZEN, 1982; DIAMOND, 1982), increasing returns to scale in production (WEIL, 1989), and even the Keynesian multiplier. Strategic complementarity also relates to speculative bubbles on stock markets which arise when some traders believe that other traders, for whatever reasons, would be willing to pay more than the asset is worth and decide to pay a high price themselves in the hope of extracting some capital gains (SHILLER, 1984). See CAMERER (1989) or SUNDER (1995) for a survey of the experimental evidence, SMITH *et al.* (1988) for a well-known experimental example.

exogenous monetary shocks will be non-neutral[16], but for other reasons than supposed.[17] New Keynesian models explain nominal rigidity from the assumption of full rationality (including rational expectations) and introduce exogenous nominal frictions into models of imperfect competition. In this experimental design there are no exogenous nominal frictions (like menu costs) and nominal rigidity arises *endogenously.*

Currency reform. The findings from this study may help to answer the following theoretical "puzzle": why do *currency reforms* not seem to have any effects on output despite the fact that they constitute nominal shocks of gigantic sizes? Currency reforms are quite frequent in high-inflation countries.[18] The typical way to proceed is to announce that from a certain day on 1 "new" currency unit will replace 1000 "old" currency units (e.g. 1 new Shekel replaced 1000 old Shekels in Israel 1985). On the other hand, relatively small monetary shocks (e.g. of a factor 0.17) seem to have dramatic effects on real activity. One reason suggested by the results of this study for this puzzle might be that in a currency reform economic agents are fully aware of the change *and* know that everyone else is aware of it, i.e. a currency reform is the prime example of an anticipated monetary shock where the assumption of common knowledge of rationality is approximately satisfied. This may be the case because it is cognitively simple to understand how the two components of the "veil of money", i.e. the "smallest nominal accounting unit-effect" and the "deflationing-effect" work. In currency reforms with a shift by a factor of 1000 these two effects are simple to understand and therefore agents do assume that everybody in the economy will understand (that everybody understands ...) and adjust their nominal variables accordingly.

---

16. "... the central issue on which Keynesian theory must stand or fall is the real effects of nominal disturbances. A central element of all Keynesian models is that nominal prices or wages do not adjust immediately. As a result, the models predict that independent monetary disturbances affect real activity. If this prediction is contracted by the data, it appears that the models would have to be abandoned rather than modified, and that the study of fluctuations would have to pursue the real-business-cycle models ..." ROMER (1996: 302).
17. There may be many other reasons why money is non-neutral. In particular, the results of the experimental study do not contest the relevance of any of the other explanations mentioned in Part I, 2.1. The results of this study do not allow to judge whether money illusion is more or less important than any of these explanations.
18. See MAS (1995) for a survey. However, the author explains why currency reforms may have real effects after all for political reasons (1995: 486): "... confiscation of currency occurs most often through the fine print of currency reforms." Confiscation may take place through a number of measures (MAS, 1995: 493) like: very short conversion periods, restricting the amounts of old currency that can be exchanged, restricting who can exchange, by establishing punitive conversion rates or by imposing costs on conversion.

In a currency reform one can easily discern nominal from real quantities (dividing by 1000 is easy). In contrast, with ordinary monetary shocks not everyone can be supposed to understand that 1 "old" currency unit shrinks to 0.83 "new" currency unit in real terms (this is the FISHER effect). In addition, deflationing (dividing) by an odd number like 0.83 is not quite that simple, and even if you can do it, do you really think everyone else in the economy can do it? Or that everyone in the economy believes that everyone can do it?

2.3. Implications for economic policy

It is long way from economic theory to useful policy advice. Problems in the "real world" come from an ever-changing environment and politicians ask for quick, simple and popular solutions. The temptation for economists is to provide recipes which may be popular but all too simple. In our opinion, the theoretical view that anticipated money does not matter in general is all too simple and has not received support in this experimental study. If the (simple) theory does not work in the simple environment (which is favorable for this theory) in the lab, it is unclear why it should work in more complex environments. Thus, policy makers (and their advisors) should be very sceptical about general claims that anticipated changes in monetary policy will never affect real economic activity.

Even so, economic policy has to refer to *some* theoretical views. Experimental economics tries to bridge the gap between simple and elegant theories to the complex real world. Even though experiments are in many respects much "closer to the real world" than the abstract general equilibrium theories, one should be aware that it is still a long way to go from an experimental result to useful policy advice. Hence, the conclusions in this paragraph necessarily remain *highly speculative*. We briefly mention possible implications for the "institutional design" of the economy, the conduct of monetary policy, and for the introduction of the Euro.

Automatic stabilizers. An obvious policy conclusion from our experimental results is that the damaging potential of monetary policy could be reduced if the degree of strategic complementarity is reduced. This could be achieved by introducing automatic stabilizers (they are also called built-in stabilizers) into the economy. Examples include unemployment insurance or progressive income taxation which both provide incentives for "countercyclical" behavior. However, as is typical for all policy advice, these measures may have negative side-effects in other respects (unemployment insurance may e.g. provide disincentives for job search).

Trade-off between unemployment and inflation. The debate about the trade-off between unemployment and inflation has recently received new attention (see e.g. the Winter 1997 issue of the Journal of Economic Perspectives). Some authors have reconsidered the much debated question of whether "inflation is greasing the wheels of the economy" (e.g. AKERLOF *et al.*, 1996; GROSHEN and SCHWEITZER, 1997). If our results can be applied to this issue, we would conclude that there would be considerable cost of (even anticipated) deflation. The results may apply because there are important strategic complementarities within labor markets (if a firm goes first with a nominal wage cut, its workers may get even more upset because of relative wage comparisons) and between wage and price setting (if a firm is reluctant to cut nominal wages because workers react adversely, it may also not want to cut its product price because this will squeeze its profits). However, it is unclear whether our results can be applied to the issue of disinflation at low (but positive) inflation rates. The experimental results suggest that high-number illusion is an important aspect of money illusion. If this finding proves to be robust one would expect pronounced asymmetric real effects of negative and positive monetary shocks (see also page 156). Hence, contractive monetary policy can cause much harm, but expansive monetary policy may stimulate the economy only slightly.

Conduct of monetary policy. Certainly, one should be careful to claim that "anomalies" observed in a laboratory environment will be equally important in naturally occurring economies, since people may be aware that they suffer from money illusion and find mechanisms to overcome this *ex ante* (e.g. by constraining oneself, see ELSTER, 1984). But does this help when dealing with monetary shocks in a strategic environment? We have shown that money illusion interacts with the strategic environment through expectations. One policy measure to reduce the degree of monetary non-neutrality would be to create trust in other people's rationality, in other people's trust of other people's rationality etc. However, it is not clear how this could be achieved. Possibly public policy measures (like information campaigns) or social processes may help to attenuate this interaction of rationality and expectations.[19] For example, economic agents may read the central governor's projection on the price

19. See FREY and EICHENBERGER (1994) who argue that social processes may transform anomalies.

level after a change in the quantity of money. But will this help any agent to perform better? This may be the case only if the central bank governor is trusted to have a "better model" of the economy than the man in the street. To predict the price level correctly, the central bank governor has to figure out how many agents are boundedly rational, how many agents think that other agents think that other agents think ... are boundedly rational. Thus, to form "rational" expectations in this context, one has to take into account the degree of (actual and supposed) irrationality in the economy.[20] In fact, if people trusted monetary authorities to have a better model of the economy, this would allow the coordination of expectations. This trust may grow over time if the central bank acquires a reputation for being able to exactly predict economic variables.

Euro. With respect to the introduction of the common European currency, two aspects related to money illusion are discussed below. The first is that money illusion may provide (an additional) rationale for the introduction of the Euro, the second is that money illusion could be an obstacle to its success.

A prominent argument for the creation of the Euro-zone with a single currency is that this would create more transparency and increased price competition. In the presence of money illusion (in the sense that people get confused when prices are presented in different currencies) creating a single currency may indeed reduce cognitive costs and increase transparency. However, whether this will be the most important gain of the Euro and whether this is worth the considerable cost of switching to the new currency by the year 2002 must remain an open question.

As explained in the last section, the results of this study may help the understanding of the theoretical puzzle why currency reforms seem to be neutral, whereas "small" shocks seem to be non-neutral. Governments and central banks seem to understand that money illusion may play an important role in currency reforms. We do not know of any currency reform with a completely uneven conversion ratio (like "1 new shilling is equal to 736.9 old shillings"). The introduction of the Euro might

---

20. BLINDER (1987): "RE [rational expectations] is not without theoretical difficulties. We all know that RE models often have multiple equilibria. More fundamentally, RE is theoretically coherent only in the context of a single agreed-upon model. In an economy in which different people hold different views of the world, the very notion lacks clarity. For example, if PAUL VOLCKER announces today that on New Year's Day he will raise M1 by 20 percent, I imagine LUCAS and I will make different revisions in our expectations for, say, real GNP in 1987. Whose expectations are 'rational'?"

provide a problematic counterexample because there will be *extremely uneven conversion rates* between the various currencies. For example, the conversion rate between Austrian Shillings and the Euro is supposed to be approximately 1 Euro : 13.78 ATS. Equivalently, 100 ATS will be approximately 7.26 Euros. Governments seem to be aware of the danger of money illusion and have started to launch information campaigns. In addition, some members of the European Union are considering to prescribe that prices have to be denoted in both currencies for an extended period of transition.

Appendices

A1. Instructions

The original instructions were in German. This section reprints a translation of the instructions used in the Nominal treatment for agents of type x. Examples refer to complements treatment (see page 179 for income tables).

General instructions for participants

You are participating in a scientific experiment which is funded by the Swiss National Science Foundation. The purpose of this experiment is to analyze decision making in experimental markets. If you read instructions carefully and take appropriate decisions, you may earn a considerable amount of money. At the end of the experiment all the money you earned will be immediately paid out in cash.

Each participant is paid SFr.15.- for showing up. During the experiment your income will not be calculated in Swiss Francs but in points. The total amount of points you collected during the experiment will be converted into Swiss Francs, by applying the following exchange rate:

10 Points = 15 centimes.

Here is a brief description of the experiment. A more detailed description is given below. All participants are in the role of firms, selling some product. In this experiment, there are two types of firms: firms of type x and firms of type y. Each firm has to choose a selling price in every period. The income you earn depends on the price you choose and on the prices all other firms choose.

During the experiment you are **not allowed to communicate with any other participant. If you have any questions, the experimenters will be glad to answer them.** If you do not follow these instructions you will be excluded from the experiment and deprived of all payments.

The following pages describe the procedures of the experiment in detail.

Detailed information for firms of type *x*

This experiment lasts 20 periods plus one trial period. You are not paid for the trial period. You should nevertheless take the trial period seriously since you may gain experience in this period. This experience helps you to take decisions in the other periods which are paid out. You are in the role of a firm, just as all other participants in this experiment. All participants are in **groups of 4**, i.e. every participant is in a group with three other firms. There are two firms of type *x* and two firms of type *y* in every group.

You are a firm of type x

Consequently, there are two other firms of type *y* and one more firm of type *x* in your group. No participant knows which persons are in his or her group. Yet, everybody knows that the group composition remains constant throughout the experiment. The decisions taken by other groups are irrelevant for your group.

In every period all firms simultaneously decide which selling price they set for the current period. Every firm has to choose an integer price from the interval

$$1 \leq \textbf{Selling price} \leq 30 .$$

How much you earn depends on the price you choose and on the average price of all other firms in your group. Independent of the type, the average price for every firm is calculated by the following formula:

Average price = (Sum of selling prices of <u>other</u> 3 firms) / 3

Consequently, the average price will be in the interval

$$1 \leq \textbf{Average price} \leq 30 .$$

The average price is rounded to the next integer number.

How to read the income table for a firm of type *x*

The **blue** income table shows your nominal income in points if you choose a specific price and a specific average price results in this period (see separate table). Your income at the end of the experiment is not based on nominal point income, but on real point income. The following relation between the two holds:

Real income = Nominal income / Average price of other firms

This formula holds for all firms. The real point income that will be paid out is rounded in every period to the next integer number.

Example:
Suppose, you choose a price of 2 and the actual average price is 4. In this case your nominal point income is 91 points. Your (rounded) real income is 23 points (= 91 / 4).

When you decide which price to choose, you do not yet know which average price will actually result in this period. The blue income table can consequently help you to calculate your real point income given your **expectation** on the average price of other firms.

Example:
Given an expectation on the average price you can read off the blue table the payoff you get when choosing different selling prices. For example, if you expect an average price of 30 and choose a price of 17, you get a real point income of 40 (= 1137 / 30). If you choose a price of 10 at this expected price, your expected nominal income is 847 points, and you will get a real payoff of 28 points (= 847 / 30).

Please note that you are in a group with one firm of type *x* and two firms of type *y*. To determine the income of the other firm of type *x*, you have to use the blue table. To determine the income of the other two firms of type *y*, you have to use the green income table. This table also shows nominal income in points. The same formula above is used to calculate real payoffs for firms of type *y*.

What the screens show

On both screens described below the current period is indicated in the upper left corner, and the upper right corner displays remaining time in seconds to decide or to view the screen.

The upper half of the **input screen** (see figure on next page) has three cells, where you can enter data into the computer.

Price decision: Enter an integer number between 1 and 30 into the first cell. You can activate this cell (as well as the other cells) by clicking into the cell with your mouse. If you want to revise your decision, you can erase the number by hitting the backspace key.

Expected average price: Enter an integer number between 1 and 30 into the second cell. This input does not affect your income and will not be known to other firms. Your payoff will be determined by the actual average price of this period. Please try to indicate an expectation that is as exact as possible since this is going to help you to take your own price decision.

Confidence: Enter an integer number from 1 to 6 to indicate how confident you are that the average price you expect (= number in the second cell) will actually result.

The numbers stand for:

1 = I am not at all confident that my expectation will be correct
2 = I am not very confident that my expectation will be correct
3 = I am not quite confident that my expectation will be correct
4 = I am quite confident that my expectation will be correct
5 = I am very confident that my expectation will be correct
6 = I am absolutely confident that my expectation will be correct

When you finished entering the numbers into the respective cells, press the **OK-button**. Once you have pressed the button, you cannot revise your decision any more for this period.

Figure 38: Input screen

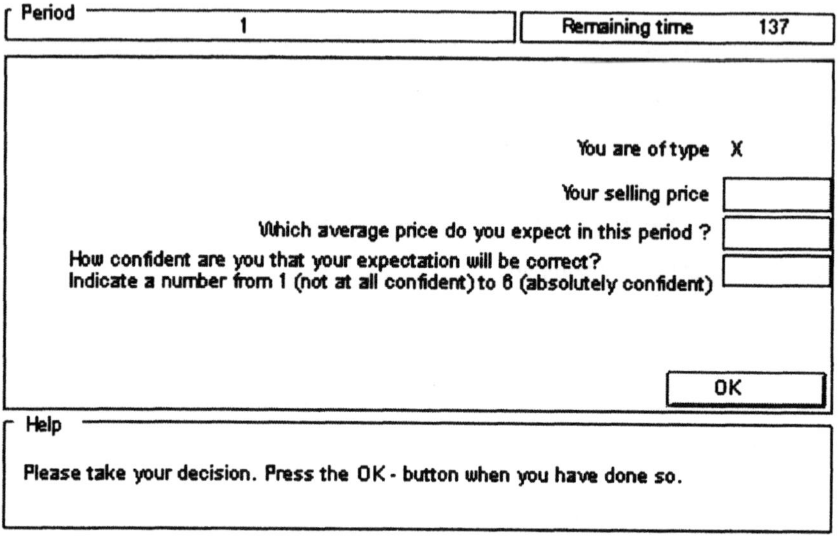

As soon as all firms have decided on their prices, the outcomes of this period will be shown in the outcome-screen.

The upper part of this screen shows the outcomes of the current period. This screen shows your decision of the current period, the average price, your real income of this period, and your total real payoff.

The lower part of this screen displays the outcomes of past periods.

Figure 39: Outcome screen

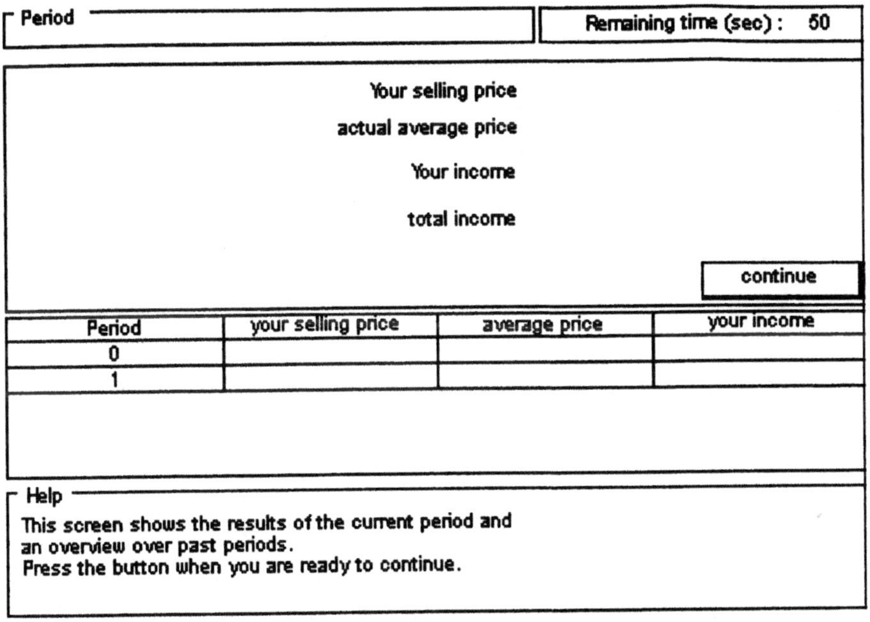

Overview: What you have to do in every period.

In every period every firm has to choose a price. Every integer price from 1 to 30 can be chosen ($1 \leq$ **Selling price** ≤ 30).

- Enter your price decision into the first cell of the input screen.
- Enter into the second cell the average you expect for this period ($1 \leq$ **Expected average price** ≤ 30).
- Enter your confidence in your price expectation into the third cell (numbers **1 to 6**).

When you have completed the three cells, press the OK-Button. The remaining time to take your decisions is shown in the upper right corner of the screen.

When all participants have taken their decisions, or when the time has elapsed, all participants are shown the outcome screen. This screen shows your decisions, actual average prices and your real payoff in points for the current and the past periods.

To take your decisions the following aids are at your disposal:

Blue income table: Helps you to estimate your expected nominal point income (You are a firm of type *x*). Your payoff is determined by your real income in points. You can calculate your real income from the nominal income (= numbers shown in the income table) by applying the following formula:

Real income = Nominal income / Average price

Green income table: Helps to estimate the nominal point income of the firms of type *y* in your group. The payoff of these firms are also determined by their real point income. To calculate the real income of firms of type *y*, you also apply the formula above.

Outcome screen: Displays your selling price, the actual average price and your real income for the present and the past periods.

Do you have any questions?

Control questions

You have to answer all of the following questions. If you do not answer a question, you will be excluded from the experiment and all payments. Wrong answers do not have any consequences. If you have any questions, please ask us.

1. Please indicate an expectation for the average price of other firms from 1 to 30.

   Expected average price ....

2. Please indicate a selling price from 1 to 30.

   Selling price ....

3. What is your expected nominal income in points at the prices you indicated in 1) and 2)?

   Your nominal income ....

4. What is your expected real income in points at the prices you indicated in 1) and 2)?

   Your real income ....

5. Suppose you choose a price of 1. The other firm of type x chooses a price of 30. The first firm of type y chooses a price of 7 and the second firm of type y chooses a price of 23.

a) What is your average price at the (fictitious) prices? ....

What is your nominal income? ....

What is your real income? ....

b) What is the average price of the other firm of type x? ....

What is the nominal income of this firm? ....

What is the real income of this firm? ....

c) What is the average price of the first firm of type y? ....

What is the nominal income of this firm? ....

What is the real income of this firm? ....

d) What is the average price of the second firm of type y?............................

What is the nominal income of this firm? ....

What is the real income of this firm? ....

A2. Income tables

The following pages show income tables that were distributed to experimental subjects.

Complements / Nominal, before shock, type x

Average price of other firms

selling price	1	2	3	4	5	6	7	8	9	10	11	12	13	14	15	16	17	18	19	20	21	22	23	24	25	26	27	28	29	30
1	18	40	68	102	143	191	243	298	351	397	433	455	464	462	453	439	423	410	403	407	424	441	442	429	409	385	360	336	314	293
2	16	35	60	91	128	172	222	278	335	390	437	473	495	503	500	490	476	463	456	461	480	499	500	486	462	434	405	377	351	326
3	14	31	53	80	113	153	200	254	312	372	429	478	514	535	543	540	530	520	514	520	542	563	565	549	523	491	458	425	394	365
4	12	28	47	71	100	136	178	228	285	346	409	468	518	554	576	585	583	577	574	583	608	632	636	620	591	556	518	480	444	410
5	11	24	42	63	88	120	158	203	256	316	380	445	506	558	596	619	629	632	634	647	675	702	709	695	666	628	586	543	501	462
6	10	22	37	56	78	106	140	180	229	284	347	413	481	544	597	637	663	677	689	707	738	769	781	772	745	707	662	614	567	522
7	9	20	33	49	69	94	123	159	202	253	312	377	446	516	581	636	679	708	731	757	792	826	845	844	825	790	745	694	641	590
8	8	18	30	44	62	83	109	141	179	224	278	339	406	478	550	617	674	720	756	789	827	865	893	905	898	872	831	780	724	668
9	7	16	27	40	55	74	97	125	158	198	246	301	365	435	508	582	650	709	758	800	840	880	918	946	956	945	915	869	814	754
10	7	14	24	36	49	66	86	111	140	175	217	267	325	390	462	536	610	679	738	786	827	868	915	960	991	1002	990	956	906	847
11	6	13	22	32	45	59	77	99	124	155	192	236	287	347	414	486	561	633	698	751	792	832	885	945	998	1034	1046	1032	996	942
12	6	12	20	29	40	53	69	88	111	138	170	209	254	307	368	435	507	579	645	699	738	778	834	903	975	1035	1075	1088	1073	1034
13	5	11	18	27	36	48	62	79	99	123	151	185	225	271	325	386	453	521	586	638	675	711	768	842	925	1006	1072	1115	1129	1113
14	5	10	17	24	33	44	56	71	89	110	134	164	199	240	288	342	402	465	525	574	608	641	695	770	858	950	1038	1109	1155	1170
15	4	9	15	22	30	40	51	64	80	98	120	146	177	213	254	302	355	412	467	512	542	572	622	693	780	877	977	1070	1146	1194
16	4	8	14	20	28	36	46	58	72	89	108	131	157	188	225	267	314	364	413	453	480	507	552	617	700	796	900	1006	1104	1183
17	4	8	13	19	26	33	42	53	65	80	97	117	141	168	200	237	278	322	365	401	424	448	488	547	623	713	815	925	1035	1137
18	3	7	12	17	24	31	39	48	60	73	88	106	126	150	178	210	246	285	323	354	375	396	432	484	551	633	729	836	950	1065
19	3	6	11	16	22	28	36	44	54	66	80	95	114	135	160	188	219	253	286	314	332	351	382	428	487	561	647	747	858	977
20	3	6	10	15	20	26	33	41	50	60	73	87	103	122	143	167	195	225	254	278	295	311	338	378	431	495	573	663	767	882
21	3	6	10	13	19	24	30	38	46	55	66	79	93	110	129	151	175	201	227	248	262	277	301	336	382	438	506	587	680	787
22	3	6	9	13	17	22	28	35	42	51	61	72	85	100	117	136	157	180	203	221	234	247	268	299	339	388	448	518	602	698
23	3	5	9	12	16	20	26	32	39	47	56	66	78	91	106	123	142	162	182	199	210	222	240	267	302	345	397	458	532	617
24	2	5	8	11	15	20	24	30	36	43	51	61	71	83	97	112	129	147	164	179	189	199	216	239	270	307	353	406	470	546
25	2	4	8	11	14	18	23	27	34	40	48	56	66	76	89	102	117	133	148	162	171	180	195	216	242	275	315	361	417	483
26	2	4	7	10	13	17	21	26	31	37	44	52	60	70	81	93	107	121	135	147	155	163	176	195	218	247	282	322	371	428
27	2	4	7	9	13	16	20	24	29	34	41	48	56	65	75	86	98	110	123	133	141	149	160	177	198	223	253	289	331	381
28	2	4	6	9	12	15	19	23	27	32	38	45	52	60	69	79	90	101	112	122	129	136	146	161	179	202	228	260	297	340
29	2	4	6	8	11	14	18	21	26	30	36	41	48	56	64	73	83	93	103	112	118	124	134	147	163	183	207	234	267	305
30	2	4	6	8	11	13	17	20	24	28	33	39	45	52	59	67	76	86	95	103	108	114	123	135	149	167	188	212	241	274

Complements / Nominal, before shock, type y

Average price of other firms

selling price	1	2	3	4	5	6	7	8	9	10	11	12	13	14	15	16	17	18	19	20	21	22	23	24	25	26	27	28	29	30
1	9	16	22	27	31	34	37	39	41	42	44	45	47	50	53	56	58	59	60	59	59	58	58	57	56	55	54	53	52	51
2	10	18	24	29	34	37	40	42	44	46	47	49	51	53	57	60	63	63	64	63	63	62	61	60	59	58	57	56	55	54
3	11	20	27	33	37	41	44	46	48	50	51	53	55	57	61	65	67	68	68	68	67	66	65	64	63	62	60	59	58	57
4	12	22	30	36	41	44	48	50	52	54	55	57	59	62	66	70	72	73	73	73	72	71	69	68	67	65	64	63	61	60
5	14	24	33	40	45	49	53	55	57	59	60	62	64	67	71	76	78	79	79	78	77	76	74	73	71	70	68	66	65	63
6	16	28	37	45	50	55	58	61	63	64	66	67	69	73	77	82	85	86	85	84	83	81	79	78	76	74	72	70	69	67
7	18	31	42	50	56	61	64	67	69	70	72	73	76	79	84	89	92	93	92	91	89	87	85	83	81	79	77	75	73	71
8	20	36	47	56	63	68	72	74	76	78	79	80	83	86	92	97	101	101	100	99	97	94	92	89	87	84	82	80	78	76
9	23	40	54	64	71	76	80	83	84	86	87	88	91	95	101	107	110	110	109	107	105	102	99	96	93	90	88	85	83	80
10	26	46	61	72	80	86	90	92	94	95	96	97	100	104	111	117	121	121	119	117	114	110	107	104	100	97	94	91	88	86
11	29	51	69	81	90	97	101	103	105	106	107	108	110	115	122	129	133	133	131	128	124	120	116	112	108	104	101	97	94	91
12	32	58	77	92	102	109	114	116	118	118	119	120	122	127	135	143	147	147	144	140	136	131	126	121	117	113	109	105	101	98
13	35	64	87	104	116	123	128	131	132	133	133	134	136	141	150	159	163	162	159	154	149	143	138	132	127	122	117	113	109	104
14	37	70	96	116	130	140	145	148	149	149	149	149	152	157	167	177	181	180	176	170	164	157	150	144	138	132	127	122	117	112
15	39	74	105	129	146	157	164	168	169	168	168	168	170	176	187	198	203	201	196	189	181	173	165	158	151	144	138	132	126	121
16	40	78	113	141	162	176	185	189	191	191	190	189	191	198	210	223	227	225	218	209	201	191	182	173	165	157	150	143	136	130
17	39	79	117	151	177	195	207	213	216	216	214	214	216	223	237	251	256	253	245	235	223	212	201	191	181	172	163	155	148	141
18	37	78	119	157	189	213	229	239	243	244	243	242	244	252	268	283	289	285	275	263	250	236	223	211	199	189	179	170	161	153
19	35	74	117	159	197	227	250	264	272	274	274	274	276	285	303	321	327	322	311	296	280	264	249	234	221	208	196	186	176	167
20	32	69	112	156	199	236	266	287	300	306	308	309	312	323	343	363	370	364	351	334	315	296	278	261	245	230	216	204	192	182
21	29	63	104	149	195	239	277	306	326	337	343	346	351	364	387	409	418	412	397	378	356	334	312	292	273	256	240	225	211	199
22	25	57	95	139	186	234	279	317	346	365	377	385	393	408	434	459	470	465	449	427	403	377	352	328	306	285	266	249	233	219
23	23	50	85	126	173	223	273	319	357	387	407	421	434	453	482	511	524	521	506	483	456	426	397	369	343	319	297	276	258	241
24	20	45	76	113	158	206	259	311	358	398	429	452	471	495	527	559	578	579	566	544	515	483	450	417	387	358	332	308	287	267
25	18	40	67	101	142	188	240	295	349	398	439	472	500	530	565	601	625	633	627	608	579	545	509	472	437	403	373	345	320	297
26	16	35	60	89	126	169	219	273	330	386	436	480	517	552	591	629	660	679	682	671	646	613	574	534	494	456	420	387	358	331
27	14	31	53	79	112	151	196	248	305	364	421	473	519	560	600	640	678	709	727	727	712	683	645	603	559	516	475	437	402	370
28	12	28	47	70	99	134	175	223	277	335	395	452	505	550	591	631	676	720	753	771	770	751	719	677	631	583	537	493	453	416
29	11	24	41	62	87	118	155	198	248	304	363	422	478	526	565	605	654	708	758	795	814	811	790	754	708	658	607	558	512	469
30	10	22	37	55	77	104	137	175	221	272	328	386	442	490	527	565	616	677	741	796	836	855	851	827	788	739	686	631	579	530

Complements / Nominal, after shock, type x

Average price of other firms

selling price	1	2	3	4	5	6	7	8	9	10	11	12	13	14	15	16	17	18	19	20	21	22	23	24	25	26	27	28	29	30
1	18	51	104	159	181	173	181	183	153	122	98	80	66	56	48	42	37	33	30	27	25	23	21	20	18	17	16	15	14	13
2	12	35	76	137	199	226	246	257	221	174	135	107	86	71	60	51	45	39	35	31	28	26	24	22	20	19	18	17	16	14
3	9	24	53	100	169	236	280	315	305	251	193	148	116	93	76	64	54	47	41	37	33	30	27	25	23	21	19	18	17	16
4	7	18	37	69	123	193	246	301	358	344	279	212	161	124	99	81	67	57	49	43	38	34	31	28	26	23	22	20	19	18
5	5	13	27	49	85	137	181	231	326	398	382	306	230	173	133	105	85	71	60	52	45	40	35	32	29	26	24	22	21	19
6	4	10	20	35	59	95	125	161	243	355	437	417	332	247	185	141	111	90	74	63	54	47	41	37	33	30	27	25	23	21
7	3	8	15	26	43	67	87	112	169	262	386	474	451	356	264	196	149	117	94	78	66	56	49	43	38	34	31	28	26	24
8	3	6	12	20	32	49	63	80	118	182	284	417	511	485	381	281	208	157	123	99	81	68	58	51	45	40	35	32	29	27
9	2	5	10	16	25	37	47	59	84	127	196	305	447	548	518	405	297	219	165	129	103	85	71	61	53	46	41	37	33	30
10	2	4	8	13	20	29	36	45	63	91	137	212	328	478	583	550	429	313	230	173	135	108	88	74	63	54	48	42	38	34
11	2	4	7	11	16	23	29	35	48	68	99	148	227	349	508	619	581	452	329	241	181	140	112	92	77	65	56	49	44	39
12	1	3	6	9	13	18	23	28	38	52	74	107	159	242	371	538	653	612	475	345	251	188	146	116	95	79	67	58	51	44
13	1	2	5	8	11	15	19	23	30	41	57	80	114	169	257	393	567	687	643	497	360	262	196	151	120	98	82	69	60	52
14	1	2	4	7	9	13	16	19	25	33	45	61	85	122	180	272	414	596	721	673	519	375	273	203	157	125	102	85	72	62
15	1	2	4	5	8	11	13	16	21	27	36	48	66	91	130	190	286	435	625	754	703	541	390	283	211	163	129	105	87	74
16	1	2	3	5	7	9	12	14	18	23	30	39	52	70	97	137	200	301	455	653	787	732	563	406	293	219	168	133	108	90
17	1	2	3	4	6	8	10	11	15	19	25	32	42	55	75	103	145	211	315	476	681	820	762	585	421	304	226	174	137	112
18	1	2	3	4	5	7	9	10	13	17	21	27	34	45	59	79	108	152	221	329	496	709	852	791	607	435	314	233	179	142
19	1	1	2	3	5	6	8	9	11	14	18	23	29	37	48	62	83	114	159	231	343	516	737	884	820	628	450	325	241	185
20	1	1	2	3	4	6	7	8	10	13	16	20	24	31	39	50	66	88	119	167	240	357	536	764	916	849	649	465	335	248
21	1	1	2	3	4	5	6	7	9	11	14	17	21	26	33	42	53	69	92	124	174	250	371	556	792	948	878	671	480	345
22	0	1	2	3	4	5	6	7	8	10	12	15	18	22	28	35	44	56	73	96	130	181	260	385	576	818	980	906	692	494
23	0	0	2	2	3	4	5	6	7	9	11	13	16	19	24	29	37	46	59	76	100	135	188	270	399	596	846	1012	935	713
24	0	0	1	2	3	4	5	5	6	8	10	12	14	17	21	25	31	39	48	61	79	104	140	195	279	412	616	873	1044	964
25	0	0	1	1	3	3	4	5	6	7	9	10	13	15	18	22	27	33	40	50	64	82	108	146	202	289	426	635	901	1075
26	0	0	1	1	2	3	4	4	5	6	8	9	11	13	16	19	23	28	34	42	53	67	86	112	151	209	298	440	655	928
27	0	0	1	1	2	2	3	4	5	6	7	8	10	12	14	17	20	24	29	36	44	54	69	89	116	156	216	308	453	674
28	0	0	1	1	2	2	3	3	5	5	6	8	9	11	13	15	18	21	26	31	37	46	57	72	92	120	161	223	317	467
29	0	0	1	1	2	2	3	3	4	4	6	7	8	10	11	13	16	19	22	27	32	39	48	59	74	95	124	166	230	327
30	0	0	1	1	2	2	3	3	4	4	5	6	8	9	10	12	14	17	20	23	28	33	40	49	61	77	98	128	172	236

Complements / Nominal, after shock, type y

Average price of other firms

selling price	1	2	3	4	5	6	7	8	9	10	11	12	13	14	15	16	17	18	19	20	21	22	23	24	25	26	27	28	29	30
1	9	14	16	17	20	23	22	21	20	19	18	17	16	15	14	13	13	12	12	11	11	10	10	9	9	9	8	8	8	8
2	12	18	21	22	26	29	28	26	24	22	21	19	18	17	16	15	14	13	13	12	12	11	11	10	10	9	9	9	8	8
3	18	25	28	29	34	37	35	32	29	27	25	23	21	19	18	17	16	15	14	13	13	12	11	11	10	10	10	9	9	9
4	26	36	39	40	45	49	45	40	36	33	29	27	24	22	21	19	18	17	16	15	14	13	13	11	11	11	10	10	10	9
5	35	52	56	56	62	67	60	53	46	40	36	32	29	26	24	22	20	19	18	16	15	15	14	13	12	12	11	11	10	10
6	40	71	81	81	89	95	83	70	60	51	44	39	34	31	28	25	23	21	20	18	17	16	15	14	14	12	12	12	11	11
7	35	80	109	115	129	137	119	97	80	66	56	48	42	37	33	30	27	24	22	21	19	18	17	16	15	14	13	13	12	11
8	25	68	119	151	176	193	172	139	111	89	73	61	52	45	39	35	31	28	26	24	22	20	19	17	16	15	14	13	13	12
9	18	50	102	157	200	236	237	201	158	123	98	79	66	56	48	42	37	33	30	27	25	23	21	19	18	17	16	15	14	13
10	12	34	74	129	176	226	279	275	229	177	136	107	86	71	59	51	44	39	35	31	28	26	23	22	20	19	17	16	15	14
11	9	24	51	92	129	173	253	318	312	255	195	148	115	92	75	63	54	46	41	36	32	29	27	24	22	21	19	18	17	16
12	7	18	36	63	89	121	189	284	357	348	281	212	160	124	98	80	67	57	49	43	38	34	30	28	25	23	21	20	19	17
13	5	13	26	45	62	84	131	210	316	395	382	305	229	172	132	104	84	70	59	51	45	39	35	32	29	26	24	22	21	19
14	4	10	19	33	45	60	91	145	232	347	433	416	331	246	183	140	110	89	74	62	53	46	41	36	33	30	27	25	23	21
15	3	8	15	25	34	44	66	102	161	254	379	470	449	355	262	194	148	116	93	77	65	56	48	43	38	34	31	28	26	24
16	3	6	12	19	26	34	49	73	112	176	277	410	506	481	378	278	206	156	122	98	81	68	58	50	44	39	35	32	29	26
17	2	5	10	15	20	26	37	54	81	123	192	300	440	541	513	402	294	217	164	127	102	84	70	60	52	46	41	36	33	30
18	2	4	8	12	16	21	29	42	60	89	134	207	322	471	576	544	425	310	227	171	133	107	87	73	62	54	47	41	37	34
19	2	4	7	10	14	17	24	33	46	66	97	145	223	344	500	611	575	448	326	238	179	139	111	91	76	64	56	49	43	39
20	1	3	6	9	11	14	19	27	37	51	72	105	156	238	365	530	645	606	470	341	249	186	144	115	94	79	67	58	50	44
21	1	2	5	7	10	12	16	22	30	40	56	78	112	166	253	386	559	679	636	492	357	259	194	150	119	97	81	69	59	52
22	1	2	4	6	8	10	14	18	24	32	44	60	84	120	177	267	407	588	712	666	514	372	270	202	156	124	101	84	71	61
23	1	2	4	5	7	8	12	15	20	27	35	48	65	90	128	187	282	428	616	745	695	536	387	280	209	161	128	104	87	73
24	1	2	3	5	6	8	10	13	17	22	29	38	51	69	95	135	197	296	449	645	778	725	558	402	291	217	167	132	108	89
25	1	2	3	4	6	7	9	11	15	19	24	32	41	55	73	101	143	207	311	469	673	811	754	580	417	301	224	172	136	111
26	1	2	3	4	5	6	8	10	13	16	21	26	34	44	58	78	107	150	218	325	490	701	843	783	601	432	312	232	178	141
27	1	1	2	3	4	5	7	9	11	14	18	22	28	36	47	61	82	112	157	227	339	510	728	875	813	623	447	322	239	183
28	1	1	2	3	4	5	6	8	10	12	15	19	24	30	39	50	65	86	117	164	237	353	530	755	908	842	644	461	332	247
29	1	1	2	3	4	4	6	7	9	11	13	17	21	26	32	41	52	68	90	123	171	247	367	550	784	940	870	666	477	343
30	0	1	2	3	3	4	5	6	8	10	12	15	18	22	27	34	43	55	72	95	128	179	257	381	570	811	972	899	687	491

Complements / Semi-Real, before shock, type *x*

Average price of other firms

selling price	1	2	3	4	5	6	7	8	9	10	11	12	13	14	15	16	17	18	19	20	21	22	23	24	25	26	27	28	29	30
1	18	20	23	26	29	32	35	37	39	40	39	38	36	33	30	27	25	23	21	20	20	20	19	18	16	15	13	12	11	10
2	16	18	20	23	26	29	32	35	37	39	40	39	38	36	33	31	28	26	24	23	23	23	22	20	18	17	15	13	12	11
3	14	16	18	20	23	26	29	32	35	37	39	40	40	38	36	34	31	29	27	26	26	26	25	23	21	19	17	15	14	12
4	12	14	16	18	20	23	25	29	32	35	37	39	40	40	38	37	34	32	30	29	29	29	28	26	24	21	19	17	15	14
5	11	12	14	16	18	20	23	25	28	32	35	37	39	40	40	39	37	35	33	32	32	32	31	29	27	24	22	19	17	15
6	10	11	12	14	16	18	20	23	25	28	32	34	37	39	40	40	39	38	36	35	35	35	34	32	30	27	25	22	20	17
7	9	10	11	12	14	16	18	20	22	25	28	31	34	37	39	40	40	39	38	38	38	38	37	35	33	30	28	25	22	20
8	8	9	10	11	12	14	16	18	20	22	25	28	31	34	37	39	40	40	40	39	39	39	39	38	36	34	31	28	25	22
9	7	8	9	10	11	12	14	16	18	20	22	25	28	31	34	36	38	39	40	40	40	40	40	39	38	36	34	31	28	25
10	7	7	8	9	10	11	12	14	16	18	20	22	25	28	31	34	36	38	39	39	39	39	40	40	40	39	37	34	31	28
11	6	7	7	8	9	10	11	12	14	16	17	20	22	25	28	30	33	35	37	38	38	38	38	39	40	40	39	37	34	31
12	6	6	7	7	8	9	10	11	12	14	15	17	20	22	25	27	30	32	34	35	35	35	36	38	39	40	40	39	37	34
13	5	6	6	7	7	8	9	10	11	12	14	15	17	19	22	24	27	29	31	32	32	32	33	35	37	39	40	40	39	37
14	5	5	6	6	7	7	8	9	10	11	12	14	15	17	19	21	24	26	28	29	29	29	30	32	34	37	38	40	40	39
15	4	5	5	6	6	7	7	8	9	10	11	12	14	15	17	19	21	23	25	26	26	26	27	29	31	34	36	38	40	40
16	4	4	5	5	6	6	7	7	8	9	10	11	12	13	15	17	18	20	22	23	23	23	24	26	28	31	33	36	38	39
17	4	4	4	5	5	6	6	7	7	8	9	10	11	12	13	15	16	18	19	20	20	20	21	23	25	27	30	33	36	38
18	3	4	4	4	5	5	6	6	7	7	8	9	10	11	12	13	14	16	17	18	18	18	19	20	22	24	27	30	33	36
19	3	3	4	4	4	5	5	6	6	7	7	8	9	10	11	12	13	14	15	16	16	16	17	18	19	22	24	27	30	33
20	3	3	3	4	4	4	5	5	6	6	7	7	8	9	10	10	11	13	13	14	14	14	15	16	17	19	21	24	26	29
21	3	3	3	3	4	4	4	5	5	6	6	7	7	8	9	9	10	11	12	12	12	13	13	14	15	17	19	21	23	26
22	3	3	3	3	3	4	4	4	5	5	6	6	7	7	8	9	9	10	11	11	11	11	12	12	14	15	17	19	21	23
23	3	3	3	3	3	3	4	4	4	5	5	6	6	7	7	8	8	9	10	10	10	10	10	11	12	13	15	16	18	21
24	2	3	3	3	3	3	3	4	4	4	5	5	5	6	6	7	8	8	9	9	9	9	9	10	11	12	13	15	16	18
25	2	2	3	3	3	3	3	3	4	4	4	5	5	5	6	6	7	7	8	8	8	8	8	9	10	11	12	13	14	16
26	2	2	2	3	3	3	3	3	3	4	4	4	5	5	5	6	6	7	7	7	7	7	8	8	9	10	10	12	13	14
27	2	2	2	2	3	3	3	3	3	3	4	4	4	5	5	5	6	6	6	7	7	7	7	7	8	9	9	10	11	13
28	2	2	2	2	2	3	3	3	3	3	3	4	4	4	5	5	5	6	6	6	6	6	6	7	7	8	8	9	10	11
29	2	2	2	2	2	2	3	3	3	3	3	3	4	4	4	5	5	5	5	6	6	6	6	6	7	7	8	8	9	10
30	2	2	2	2	2	2	2	3	3	3	3	3	3	4	4	4	4	5	5	5	5	5	5	6	6	6	7	8	8	9

Complements / Semi-Real, before shock, type *y*

Average price of other firms

selling price	1	2	3	4	5	6	7	8	9	10	11	12	13	14	15	16	17	18	19	20	21	22	23	24	25	26	27	28	29	30
1	9	8	7	7	6	6	5	5	5	4	4	4	4	4	4	4	3	3	3	3	3	3	3	2	2	2	2	2	2	2
2	10	9	8	7	7	6	6	5	5	5	4	4	4	4	4	4	4	4	3	3	3	3	3	3	2	2	2	2	2	2
3	11	10	9	8	7	7	6	6	5	5	5	4	4	4	4	4	4	4	4	3	3	3	3	3	3	2	2	2	2	2
4	12	11	10	9	8	7	7	6	6	5	5	5	5	4	4	4	4	4	4	4	3	3	3	3	3	3	2	2	2	2
5	14	12	11	10	9	8	8	7	6	6	5	5	5	5	5	5	5	4	4	4	4	3	3	3	3	3	3	2	2	2
6	16	14	12	11	10	9	8	8	7	6	6	6	5	5	5	5	5	5	4	4	4	4	3	3	3	3	3	3	2	2
7	18	16	14	13	11	10	9	8	8	7	7	6	6	6	6	6	5	5	5	5	4	4	4	3	3	3	3	3	3	2
8	20	18	16	14	13	11	10	9	8	8	7	7	6	6	6	6	6	6	5	5	5	4	4	4	3	3	3	3	3	3
9	23	20	18	16	14	13	11	10	9	9	8	7	7	7	7	7	6	6	6	5	5	5	4	4	4	3	3	3	3	3
10	26	23	20	18	16	14	13	12	10	10	9	8	8	7	7	7	7	7	6	6	5	5	5	4	4	4	3	3	3	3
11	29	26	23	20	18	16	14	13	12	11	10	9	8	8	8	8	8	7	7	6	6	5	5	5	4	4	4	3	3	3
12	32	29	26	23	20	18	16	15	13	12	11	10	9	9	9	9	9	8	8	7	6	6	5	5	5	4	4	4	3	3
13	35	32	29	26	23	21	18	16	15	13	12	11	10	10	10	10	10	9	8	8	7	7	6	6	5	5	4	4	4	3
14	37	35	32	29	26	23	21	19	17	15	14	12	12	11	11	11	11	10	9	9	8	7	7	6	6	5	5	4	4	4
15	39	37	35	32	29	26	23	21	19	17	15	14	13	13	12	12	12	11	10	9	9	8	7	7	6	6	5	5	4	4
16	40	39	38	35	32	29	26	24	21	19	17	16	15	14	14	14	13	13	11	10	10	9	8	7	7	6	6	5	5	4
17	39	40	39	38	35	33	30	27	24	22	19	18	17	16	16	16	15	14	13	12	11	10	9	8	7	7	6	6	5	5
18	37	39	40	39	38	36	33	30	27	24	22	20	19	18	18	18	17	16	14	13	12	11	10	9	8	7	7	6	6	5
19	35	37	39	40	39	38	36	33	30	27	25	23	21	20	20	20	19	18	16	15	13	12	11	10	9	8	7	7	6	6
20	32	35	37	39	40	39	38	36	33	31	28	26	24	23	23	23	22	20	18	17	15	13	12	11	10	9	8	7	7	6
21	29	32	35	37	39	40	40	38	36	34	31	29	27	26	26	26	25	23	21	19	17	15	14	12	11	10	9	8	7	7
22	25	29	32	35	37	39	40	40	38	37	34	32	30	29	29	29	28	26	24	21	19	17	15	14	12	11	10	9	8	7
23	23	25	28	32	35	37	39	40	40	39	37	35	33	32	32	32	31	29	27	24	22	19	17	15	14	12	11	10	9	8
24	20	23	25	28	32	34	37	39	40	40	39	38	36	35	35	35	34	32	30	27	25	22	20	17	15	14	12	11	10	9
25	18	20	22	25	28	31	34	37	39	40	40	39	38	38	38	38	37	35	33	30	28	25	22	20	17	16	14	12	11	10
26	16	18	20	22	25	28	31	34	37	39	40	40	40	39	39	39	39	38	36	34	31	28	25	22	20	18	16	14	12	11
27	14	16	18	20	22	25	28	31	34	36	38	39	40	40	40	40	40	39	38	36	34	31	28	25	22	20	18	16	14	12
28	12	14	16	18	20	22	25	28	31	34	36	38	39	39	39	39	40	40	40	39	37	34	31	28	25	22	20	18	16	14
29	11	12	14	16	17	20	22	25	28	30	33	35	37	38	38	38	38	39	40	40	39	37	34	31	28	25	22	20	18	16
30	10	11	12	14	15	17	20	22	25	27	30	32	34	35	35	35	36	38	39	40	40	39	37	34	32	28	25	23	20	18

Complements / Semi-Real, after shock, type *x*

Average price of other firms

selling price	1	2	3	4	5	6	7	8	9	10	11	12	13	14	15	16	17	18	19	20	21	22	23	24	25	26	27	28	29	30
1	18	26	35	40	36	29	26	23	17	12	9	7	5	4	3	3	2	2	2	1	1	1	1	1	1	1	1	1	0	0
2	12	18	25	34	40	38	35	32	25	17	12	9	7	5	4	3	3	2	2	2	1	1	1	1	1	1	1	1	1	0
3	9	12	18	25	34	39	40	39	34	25	18	12	9	7	5	4	3	3	2	2	2	1	1	1	1	1	1	1	1	1
4	7	9	12	17	25	32	35	38	40	34	25	18	12	9	7	5	4	3	3	2	2	2	1	1	1	1	1	1	1	1
5	5	7	9	12	17	23	26	29	36	40	35	26	18	12	9	7	5	4	3	3	2	2	2	1	1	1	1	1	1	1
6	4	5	7	9	12	16	18	20	27	36	40	35	26	18	12	9	7	5	4	3	3	2	2	2	1	1	1	1	1	1
7	3	4	5	7	9	11	12	14	19	26	35	40	35	25	18	12	9	7	5	4	3	3	2	2	2	1	1	1	1	1
8	3	3	4	5	6	8	9	10	13	18	26	35	39	35	25	18	12	9	6	5	4	3	3	2	2	2	1	1	1	1
9	2	3	3	4	5	6	7	7	9	13	18	25	34	39	35	25	17	12	9	6	5	4	3	3	2	2	2	1	1	1
10	2	2	3	3	4	5	5	6	7	9	12	18	25	34	39	34	25	17	12	9	6	5	4	3	3	2	2	2	1	1
11	2	2	2	3	3	4	4	4	5	7	9	12	17	25	34	39	34	25	17	12	9	6	5	4	3	3	2	2	2	1
12	1	2	2	2	3	3	3	4	4	5	7	9	12	17	25	34	38	34	25	17	12	9	6	5	4	3	2	2	2	1
13	1	1	2	2	2	3	3	3	3	4	5	7	9	12	17	25	33	38	34	25	17	12	9	6	5	4	3	2	2	2
14	1	1	1	2	2	2	2	2	3	3	4	5	7	9	12	17	24	33	38	34	25	17	12	8	6	5	4	3	2	2
15	1	1	1	1	2	2	2	2	2	3	3	4	5	7	9	12	17	24	33	38	33	25	17	12	8	6	5	4	3	2
16	1	1	1	1	1	2	2	2	2	2	3	3	4	5	6	9	12	17	24	33	37	33	24	17	12	8	6	5	4	3
17	1	1	1	1	1	1	1	1	2	2	2	3	3	4	5	6	9	12	17	24	32	37	33	24	17	12	8	6	5	4
18	1	1	1	1	1	1	1	1	1	2	2	2	3	3	4	5	6	8	12	16	24	32	37	33	24	17	12	8	6	5
19	1	1	1	1	1	1	1	1	1	1	2	2	2	3	3	4	5	6	8	12	16	23	32	37	33	24	17	12	8	6
20	1	1	1	1	1	1	1	1	1	1	1	2	2	2	3	3	4	5	6	8	11	16	23	32	37	33	24	17	12	8
21	1	1	1	1	1	1	1	1	1	1	1	1	2	2	2	3	3	4	5	6	8	11	16	23	32	36	33	24	17	12
22	0	1	1	1	1	1	1	1	1	1	1	1	1	2	2	2	3	3	4	5	6	8	11	16	23	31	36	32	24	16
23	0	0	1	1	1	1	1	1	1	1	1	1	1	1	2	2	2	3	3	4	5	6	8	11	16	23	31	36	32	24
24	0	0	0	1	1	1	1	1	1	1	1	1	1	1	1	2	2	2	3	3	4	5	6	8	11	16	23	31	36	32
25	0	0	0	0	1	1	1	1	1	1	1	1	1	1	1	1	2	2	2	3	3	4	5	6	8	11	16	23	31	36
26	0	0	0	0	0	1	1	1	1	1	1	1	1	1	1	1	1	2	2	2	3	3	4	5	6	8	11	16	23	31
27	0	0	0	0	0	0	0	1	1	1	1	1	1	1	1	1	1	1	2	2	2	2	3	4	5	6	8	11	16	22
28	0	0	0	0	0	0	0	0	1	1	1	1	1	1	1	1	1	1	1	2	2	2	2	3	4	5	6	8	11	16
29	0	0	0	0	0	0	0	0	0	0	1	1	1	1	1	1	1	1	1	1	2	2	2	2	3	4	5	6	8	11
30	0	0	0	0	0	0	0	0	0	0	0	1	1	1	1	1	1	1	1	1	1	2	2	2	2	3	4	5	6	8

Complements / Semi-Real, after shock, type *y*

Average price of other firms

selling price	1	2	3	4	5	6	7	8	9	10	11	12	13	14	15	16	17	18	19	20	21	22	23	24	25	26	27	28	29	30
1	9	7	5	4	4	4	3	3	2	2	2	1	1	1	1	1	1	1	1	1	1	0	0	0	0	0	0	0	0	0
2	12	9	7	6	5	5	4	3	3	2	2	2	1	1	1	1	1	1	1	1	1	1	0	0	0	0	0	0	0	0
3	18	13	9	7	7	6	5	4	3	3	2	2	2	1	1	1	1	1	1	1	1	1	0	0	0	0	0	0	0	0
4	26	18	13	10	9	8	6	5	4	3	3	2	2	2	1	1	1	1	1	1	1	1	1	0	0	0	0	0	0	0
5	35	26	19	14	12	11	9	7	5	4	3	3	2	2	2	1	1	1	1	1	1	1	1	1	0	0	0	0	0	0
6	40	36	27	20	18	16	12	9	7	5	4	3	3	2	2	2	1	1	1	1	1	1	1	1	1	0	0	0	0	0
7	35	40	36	29	26	23	17	12	9	7	5	4	3	3	2	2	2	1	1	1	1	1	1	1	1	1	0	0	0	0
8	25	34	40	38	35	32	25	17	12	9	7	5	4	3	3	2	2	2	1	1	1	1	1	1	1	1	1	0	0	0
9	18	25	34	39	40	39	34	25	18	12	9	7	5	4	3	3	2	2	2	1	1	1	1	1	1	1	1	1	0	0
10	12	17	25	32	35	38	40	34	25	18	12	9	7	5	4	3	3	2	2	2	1	1	1	1	1	1	1	1	1	0
11	9	12	17	23	26	29	36	40	35	26	18	12	9	7	5	4	3	3	2	2	2	1	1	1	1	1	1	1	1	1
12	7	9	12	16	18	20	27	36	40	35	26	18	12	9	7	5	4	3	3	2	2	2	1	1	1	1	1	1	1	1
13	5	7	9	11	12	14	19	26	35	40	35	25	18	12	9	7	5	4	3	3	2	2	2	1	1	1	1	1	1	1
14	4	5	6	8	9	10	13	18	26	35	39	35	25	18	12	9	6	5	4	3	3	2	2	2	1	1	1	1	1	1
15	3	4	5	6	7	7	9	13	18	25	34	39	35	25	17	12	9	6	5	4	3	3	2	2	2	1	1	1	1	1
16	3	3	4	5	5	6	7	9	12	18	25	34	39	34	25	17	12	9	6	5	4	3	3	2	2	2	1	1	1	1
17	2	3	3	4	4	4	5	7	9	12	17	25	34	39	34	25	17	12	9	6	5	4	3	3	2	2	2	1	1	1
18	2	2	3	3	3	4	4	5	7	9	12	17	25	34	38	34	25	17	12	9	6	5	4	3	2	2	2	1	1	1
19	2	2	2	3	3	3	3	4	5	7	9	12	17	25	33	38	34	25	17	12	9	6	5	4	3	2	2	2	1	1
20	1	2	2	2	2	2	3	3	4	5	7	9	12	17	24	33	38	34	25	17	12	8	6	5	4	3	2	2	2	1
21	1	1	2	2	2	2	2	3	3	4	5	7	9	12	17	24	33	38	33	25	17	12	8	6	5	4	3	2	2	2
22	1	1	1	2	2	2	2	2	3	3	4	5	6	9	12	17	24	33	37	33	24	17	12	8	6	5	4	3	2	2
23	1	1	1	1	1	1	2	2	2	3	3	4	5	6	9	12	17	24	32	37	33	24	17	12	8	6	5	4	3	2
24	1	1	1	1	1	1	1	2	2	2	3	3	4	5	6	8	12	16	24	32	37	33	24	17	12	8	6	5	4	3
25	1	1	1	1	1	1	1	1	2	2	2	3	3	4	5	6	8	12	16	23	32	37	33	24	17	12	8	6	5	4
26	1	1	1	1	1	1	1	1	1	2	2	2	3	3	4	5	6	8	11	16	23	32	37	33	24	17	12	8	6	5
27	1	1	1	1	1	1	1	1	1	1	2	2	2	3	3	4	5	6	8	11	16	23	32	36	33	24	17	12	8	6
28	1	1	1	1	1	1	1	1	1	1	1	2	2	2	3	3	4	5	6	8	11	16	23	31	36	32	24	16	11	8
29	1	1	1	1	1	1	1	1	1	1	1	1	2	2	2	3	3	4	5	6	8	11	16	23	31	36	32	24	16	11
30	0	1	1	1	1	1	1	1	1	1	1	1	1	2	2	2	3	3	4	5	6	8	11	16	23	31	36	32	24	16

Complements / Pure Real, before shock, type x

Average price of other firms

selling price	3	6	9	12	15	18	21	24	27	30	33	36	39	42	45	48	51	54	57	60	63	66	69	72	75	78	81	84	87	90
3	18	26	35	40	36	29	26	23	17	12	9	7	5	4	3	3	2	2	2	1	1	1	1	1	1	1	1	1	0	0
6	12	18	25	34	40	38	35	32	25	17	12	9	7	5	4	3	3	2	2	2	1	1	1	1	1	1	1	1	1	0
9	9	12	18	25	34	39	40	39	34	25	18	12	9	7	5	4	3	3	2	2	2	1	1	1	1	1	1	1	1	1
12	7	9	12	17	25	32	35	38	40	34	25	18	12	9	7	5	4	3	3	2	2	2	1	1	1	1	1	1	1	1
15	5	7	9	12	17	23	26	29	36	40	35	26	18	12	9	7	5	4	3	3	2	2	2	1	1	1	1	1	1	1
18	4	5	7	9	12	16	18	20	27	36	40	35	26	18	12	9	7	5	4	3	3	2	2	2	1	1	1	1	1	1
21	3	4	5	7	9	11	12	14	19	26	35	40	35	25	18	12	9	7	5	4	3	3	2	2	2	1	1	1	1	1
24	3	3	4	5	6	8	9	10	13	18	26	35	39	35	25	18	12	9	6	5	4	3	3	2	2	2	1	1	1	1
27	2	3	3	4	5	6	7	7	9	13	18	25	34	39	35	25	17	12	9	6	5	4	3	3	2	2	2	1	1	1
30	2	2	3	3	4	5	5	6	7	9	12	18	25	34	39	34	25	17	12	9	6	5	4	3	3	2	2	2	1	1
33	2	2	2	3	3	4	4	4	5	7	9	12	17	25	34	39	34	25	17	12	9	6	5	4	3	3	2	2	2	1
36	1	2	2	2	3	3	3	4	4	5	7	9	12	17	25	34	38	34	25	17	12	9	6	5	4	3	2	2	2	1
39	1	1	2	2	2	3	3	3	3	4	5	7	9	12	17	25	33	38	34	25	17	12	9	6	5	4	3	2	2	2
42	1	1	1	2	2	2	2	2	3	3	4	5	7	9	12	17	24	33	38	34	25	17	12	8	6	5	4	3	2	2
45	1	1	1	1	2	2	2	2	2	3	3	4	5	7	9	12	17	24	33	38	33	25	17	12	8	6	5	4	3	2
48	1	1	1	1	1	2	2	2	2	2	3	3	4	5	6	9	12	17	24	33	37	33	24	17	12	8	6	5	4	3
51	1	1	1	1	1	1	1	1	2	2	2	3	3	4	5	6	9	12	17	24	32	37	33	24	17	12	8	6	5	4
54	1	1	1	1	1	1	1	1	1	2	2	2	3	3	4	5	6	8	12	16	24	32	37	33	24	17	12	8	6	5
57	1	1	1	1	1	1	1	1	1	1	2	2	2	3	3	4	5	6	8	12	16	23	32	37	33	24	17	12	8	6
60	1	1	1	1	1	1	1	1	1	1	1	2	2	2	3	3	4	5	6	8	11	16	23	32	37	33	24	17	12	8
63	1	1	1	1	1	1	1	1	1	1	1	1	2	2	2	3	3	4	5	6	8	11	16	23	32	36	33	24	17	12
66	0	1	1	1	1	1	1	1	1	1	1	1	1	2	2	2	3	3	4	5	6	8	11	16	23	31	36	32	24	16
69	0	0	1	1	1	1	1	1	1	1	1	1	1	1	2	2	2	3	3	4	5	6	8	11	16	23	31	36	32	24
72	0	0	0	1	1	1	1	1	1	1	1	1	1	1	1	2	2	2	3	3	4	5	6	8	11	16	23	31	36	32
75	0	0	0	0	1	1	1	1	1	1	1	1	1	1	1	1	2	2	2	3	3	4	5	6	8	11	16	23	31	36
78	0	0	0	0	0	1	1	1	1	1	1	1	1	1	1	1	1	2	2	2	3	3	4	5	6	8	11	16	23	31
81	0	0	0	0	0	0	0	1	1	1	1	1	1	1	1	1	1	1	2	2	2	2	3	4	5	6	8	11	16	22
84	0	0	0	0	0	0	0	0	1	1	1	1	1	1	1	1	1	1	1	2	2	2	2	3	4	5	6	8	11	16
87	0	0	0	0	0	0	0	0	0	0	1	1	1	1	1	1	1	1	1	1	2	2	2	2	3	4	5	6	8	11
90	0	0	0	0	0	0	0	0	0	0	0	1	1	1	1	1	1	1	1	1	1	2	2	2	2	3	4	5	6	8

Complements / Pure Real, before shock, type *y*

Average price of other firms

selling price	3	6	9	12	15	18	21	24	27	30	33	36	39	42	45	48	51	54	57	60	63	66	69	72	75	78	81	84	87	90
3	9	7	5	4	4	4	3	3	2	2	2	1	1	1	1	1	1	1	1	1	1	0	0	0	0	0	0	0	0	0
6	12	9	7	6	5	5	4	3	3	2	2	2	1	1	1	1	1	1	1	1	1	1	0	0	0	0	0	0	0	0
9	18	13	9	7	7	6	5	4	3	3	2	2	2	1	1	1	1	1	1	1	1	1	0	0	0	0	0	0	0	0
12	26	18	13	10	9	8	6	5	4	3	3	2	2	2	1	1	1	1	1	1	1	1	1	0	0	0	0	0	0	0
15	35	26	19	14	12	11	9	7	5	4	3	3	2	2	2	1	1	1	1	1	1	1	1	1	0	0	0	0	0	0
18	40	36	27	20	18	16	12	9	7	5	4	3	3	2	2	2	1	1	1	1	1	1	1	1	1	0	0	0	0	0
21	35	40	36	29	26	23	17	12	9	7	5	4	3	3	2	2	2	1	1	1	1	1	1	1	1	1	0	0	0	0
24	25	34	40	38	35	32	25	17	12	9	7	5	4	3	3	2	2	2	1	1	1	1	1	1	1	1	1	0	0	0
27	18	25	34	39	40	39	34	25	18	12	9	7	5	4	3	3	2	2	2	1	1	1	1	1	1	1	1	1	0	0
30	12	17	25	32	35	38	40	34	25	18	12	9	7	5	4	3	3	2	2	2	1	1	1	1	1	1	1	1	1	0
33	9	12	17	23	26	29	36	40	35	26	18	12	9	7	5	4	3	3	2	2	2	1	1	1	1	1	1	1	1	1
36	7	9	12	16	18	20	27	36	40	35	26	18	12	9	7	5	4	3	3	2	2	2	1	1	1	1	1	1	1	1
39	5	7	9	11	12	14	19	26	35	40	35	25	18	12	9	7	5	4	3	3	2	2	2	1	1	1	1	1	1	1
42	4	5	6	8	9	10	13	18	26	35	39	35	25	18	12	9	6	5	4	3	3	2	2	2	1	1	1	1	1	1
45	3	4	5	6	7	7	9	13	18	25	34	39	35	25	17	12	9	6	5	4	3	3	2	2	2	1	1	1	1	1
48	3	3	4	5	5	6	7	9	12	18	25	34	39	34	25	17	12	9	6	5	4	3	3	2	2	2	1	1	1	1
51	2	3	3	4	4	4	5	7	9	12	17	25	34	39	34	25	17	12	9	6	5	4	3	3	2	2	2	1	1	1
54	2	2	3	3	3	4	4	5	7	9	12	17	25	34	38	34	25	17	12	9	6	5	4	3	2	2	2	1	1	1
57	2	2	2	3	3	3	3	4	5	7	9	12	17	25	33	38	34	25	17	12	9	6	5	4	3	2	2	2	1	1
60	1	2	2	2	2	2	3	3	4	5	7	9	12	17	24	33	38	34	25	17	12	8	6	5	4	3	2	2	2	1
63	1	1	2	2	2	2	2	3	3	4	5	7	9	12	17	24	33	38	33	25	17	12	8	6	5	4	3	2	2	2
66	1	1	1	2	2	2	2	2	3	3	4	5	6	9	12	17	24	33	37	33	24	17	12	8	6	5	4	3	2	2
69	1	1	1	1	1	1	2	2	2	3	3	4	5	6	9	12	17	24	32	37	33	24	17	12	8	6	5	4	3	2
72	1	1	1	1	1	1	1	2	2	2	3	3	4	5	6	8	12	16	24	32	37	33	24	17	12	8	6	5	4	3
75	1	1	1	1	1	1	1	1	2	2	2	3	3	4	5	6	8	12	16	23	32	37	33	24	17	12	8	6	5	4
78	1	1	1	1	1	1	1	1	1	2	2	2	3	3	4	5	6	8	11	16	23	32	37	33	24	17	12	8	6	5
81	1	1	1	1	1	1	1	1	1	1	2	2	2	3	3	4	5	6	8	11	16	23	32	36	33	24	17	12	8	6
84	1	1	1	1	1	1	1	1	1	1	1	2	2	2	3	3	4	5	6	8	11	16	23	31	36	32	24	16	11	8
87	1	1	1	1	1	1	1	1	1	1	1	1	2	2	2	3	3	4	5	6	8	11	16	23	31	36	32	24	16	11
90	0	1	1	1	1	1	1	1	1	1	1	1	1	2	2	2	3	3	4	5	6	8	11	16	23	31	36	32	24	16

Complements / Pure Real, after shock, type *x*

Average price of other firms

selling price	1	2	3	4	5	6	7	8	9	10	11	12	13	14	15	16	17	18	19	20	21	22	23	24	25	26	27	28	29	30
1	18	26	35	40	36	29	26	23	17	12	9	7	5	4	3	3	2	2	2	1	1	1	1	1	1	1	1	1	0	0
2	12	18	25	34	40	38	35	32	25	17	12	9	7	5	4	3	3	2	2	2	1	1	1	1	1	1	1	1	1	0
3	9	12	18	25	34	39	40	39	34	25	18	12	9	7	5	4	3	3	2	2	2	1	1	1	1	1	1	1	1	1
4	7	9	12	17	25	32	35	38	40	34	25	18	12	9	7	5	4	3	3	2	2	2	1	1	1	1	1	1	1	1
5	5	7	9	12	17	23	26	29	36	40	35	26	18	12	9	7	5	4	3	3	2	2	2	1	1	1	1	1	1	1
6	4	5	7	9	12	16	18	20	27	36	40	35	26	18	12	9	7	5	4	3	3	2	2	2	1	1	1	1	1	1
7	3	4	5	7	9	11	12	14	19	26	35	40	35	25	18	12	9	7	5	4	3	3	2	2	2	1	1	1	1	1
8	3	3	4	5	6	8	9	10	13	18	26	35	39	35	25	18	12	9	6	5	4	3	3	2	2	2	1	1	1	1
9	2	3	3	4	5	6	7	7	9	13	18	25	34	39	35	25	17	12	9	6	5	4	3	3	2	2	2	1	1	1
10	2	2	3	3	4	5	5	6	7	9	12	18	25	34	39	34	25	17	12	9	6	5	4	3	3	2	2	2	1	1
11	2	2	2	3	3	4	4	4	5	7	9	12	17	25	34	39	34	25	17	12	9	6	5	4	3	3	2	2	2	1
12	1	2	2	2	3	3	3	4	4	5	7	9	12	17	25	34	38	34	25	17	12	9	6	5	4	3	2	2	2	1
13	1	1	2	2	2	3	3	3	3	4	5	7	9	12	17	25	33	38	34	25	17	12	9	6	5	4	3	2	2	2
14	1	1	1	2	2	2	2	2	3	3	4	5	7	9	12	17	24	33	38	34	25	17	12	8	6	5	4	3	2	2
15	1	1	1	1	2	2	2	2	2	3	3	4	5	7	9	12	17	24	33	38	33	25	17	12	8	6	5	4	3	2
16	1	1	1	1	1	2	2	2	2	2	3	3	4	5	6	9	12	17	24	33	37	33	24	17	12	8	6	5	4	3
17	1	1	1	1	1	1	1	1	2	2	2	3	3	4	5	6	9	12	17	24	32	37	33	24	17	12	8	6	5	4
18	1	1	1	1	1	1	1	1	1	2	2	2	3	3	4	5	6	8	12	16	24	32	37	33	24	17	12	8	6	5
19	1	1	1	1	1	1	1	1	1	1	2	2	2	3	3	4	5	6	8	12	16	23	32	37	33	24	17	12	8	6
20	1	1	1	1	1	1	1	1	1	1	1	2	2	2	3	3	4	5	6	8	11	16	23	32	37	33	24	17	12	8
21	1	1	1	1	1	1	1	1	1	1	1	1	2	2	2	3	3	4	5	6	8	11	16	23	32	36	33	24	17	12
22	0	1	1	1	1	1	1	1	1	1	1	1	1	2	2	2	3	3	4	5	6	8	11	16	23	31	36	32	24	16
23	0	0	1	1	1	1	1	1	1	1	1	1	1	1	2	2	2	3	3	4	5	6	8	11	16	23	31	36	32	24
24	0	0	0	1	1	1	1	1	1	1	1	1	1	1	1	2	2	2	3	3	4	5	6	8	11	16	23	31	36	32
25	0	0	0	0	1	1	1	1	1	1	1	1	1	1	1	1	2	2	2	3	3	4	5	6	8	11	16	23	31	36
26	0	0	0	0	0	1	1	1	1	1	1	1	1	1	1	1	1	2	2	2	3	3	4	5	6	8	11	16	23	31
27	0	0	0	0	0	0	0	1	1	1	1	1	1	1	1	1	1	1	2	2	2	2	3	4	5	6	8	11	16	22
28	0	0	0	0	0	0	0	0	1	1	1	1	1	1	1	1	1	1	1	2	2	2	2	3	4	5	6	8	11	16
29	0	0	0	0	0	0	0	0	0	0	1	1	1	1	1	1	1	1	1	1	2	2	2	2	3	4	5	6	8	11
30	0	0	0	0	0	0	0	0	0	0	0	1	1	1	1	1	1	1	1	1	1	2	2	2	2	3	4	5	6	8

Complements / Pure Real, after shock, type *y*

Average price of other firms

selling price	1	2	3	4	5	6	7	8	9	10	11	12	13	14	15	16	17	18	19	20	21	22	23	24	25	26	27	28	29	30
1	9	7	5	4	4	4	3	3	2	2	2	1	1	1	1	1	1	1	1	1	1	0	0	0	0	0	0	0	0	0
2	12	9	7	6	5	5	4	3	3	2	2	2	1	1	1	1	1	1	1	1	1	1	0	0	0	0	0	0	0	0
3	18	13	9	7	7	6	5	4	3	3	2	2	2	1	1	1	1	1	1	1	1	1	0	0	0	0	0	0	0	0
4	26	18	13	10	9	8	6	5	4	3	3	2	2	2	1	1	1	1	1	1	1	1	1	0	0	0	0	0	0	0
5	35	26	19	14	12	11	9	7	5	4	3	3	2	2	2	1	1	1	1	1	1	1	1	1	0	0	0	0	0	0
6	40	36	27	20	18	16	12	9	7	5	4	3	3	2	2	2	1	1	1	1	1	1	1	1	1	0	0	0	0	0
7	35	40	36	29	26	23	17	12	9	7	5	4	3	3	2	2	2	1	1	1	1	1	1	1	1	1	0	0	0	0
8	25	34	40	38	35	32	25	17	12	9	7	5	4	3	3	2	2	2	1	1	1	1	1	1	1	1	1	0	0	0
9	18	25	34	39	40	39	34	25	18	12	9	7	5	4	3	3	2	2	2	1	1	1	1	1	1	1	1	1	0	0
10	12	17	25	32	35	38	40	34	25	18	12	9	7	5	4	3	3	2	2	2	1	1	1	1	1	1	1	1	1	0
11	9	12	17	23	26	29	36	40	35	26	18	12	9	7	5	4	3	3	2	2	2	1	1	1	1	1	1	1	1	1
12	7	9	12	16	18	20	27	36	40	35	26	18	12	9	7	5	4	3	3	2	2	2	1	1	1	1	1	1	1	1
13	5	7	9	11	12	14	19	26	35	40	35	25	18	12	9	7	5	4	3	3	2	2	2	1	1	1	1	1	1	1
14	4	5	6	8	9	10	13	18	26	35	39	35	25	18	12	9	6	5	4	3	3	2	2	2	1	1	1	1	1	1
15	3	4	5	6	7	7	9	13	18	25	34	39	35	25	17	12	9	6	5	4	3	3	2	2	2	1	1	1	1	1
16	3	3	4	5	5	6	7	9	12	18	25	34	39	34	25	17	12	9	6	5	4	3	3	2	2	2	1	1	1	1
17	2	3	3	4	4	4	5	7	9	12	17	25	34	39	34	25	17	12	9	6	5	4	3	3	2	2	2	1	1	1
18	2	2	3	3	3	4	4	5	7	9	12	17	25	34	38	34	25	17	12	9	6	5	4	3	2	2	2	1	1	1
19	2	2	2	3	3	3	3	4	5	7	9	12	17	25	33	38	34	25	17	12	9	6	5	4	3	2	2	2	1	1
20	1	2	2	2	2	2	3	3	4	5	7	9	12	17	24	33	38	34	25	17	12	8	6	5	4	3	2	2	2	1
21	1	1	2	2	2	2	2	3	3	4	5	7	9	12	17	24	33	38	33	25	17	12	8	6	5	4	3	2	2	2
22	1	1	1	2	2	2	2	2	3	3	4	5	6	9	12	17	24	33	37	33	24	17	12	8	6	5	4	3	2	2
23	1	1	1	1	1	1	2	2	2	3	3	4	5	6	9	12	17	24	32	37	33	24	17	12	8	6	5	4	3	2
24	1	1	1	1	1	1	1	2	2	2	3	3	4	5	6	8	12	16	24	32	37	33	24	17	12	8	6	5	4	3
25	1	1	1	1	1	1	1	1	2	2	2	3	3	4	5	6	8	12	16	23	32	37	33	24	17	12	8	6	5	4
26	1	1	1	1	1	1	1	1	1	2	2	2	3	3	4	5	6	8	11	16	23	32	37	33	24	17	12	8	6	5
27	1	1	1	1	1	1	1	1	1	1	2	2	2	3	3	4	5	6	8	11	16	23	32	36	33	24	17	12	8	6
28	1	1	1	1	1	1	1	1	1	1	1	2	2	2	3	3	4	5	6	8	11	16	23	31	36	32	24	16	11	8
29	1	1	1	1	1	1	1	1	1	1	1	1	2	2	2	3	3	4	5	6	8	11	16	23	31	36	32	24	16	11
30	0	1	1	1	1	1	1	1	1	1	1	1	1	2	2	2	3	3	4	5	6	8	11	16	23	31	36	32	24	16

Substitutes / Nominal, before shock, type *x*

Average price of other firms

selling price	1	2	3	4	5	6	7	8	9	10	11	12	13	14	15	16	17	18	19	20	21	22	23	24	25	26	27	28	29	30
1	4	8	13	19	26	33	42	53	65	80	97	117	141	168	200	237	278	322	365	401	424	448	488	547	623	713	815	925	1035	1137
2	4	8	14	20	28	36	46	58	72	89	108	131	157	188	225	267	314	364	413	453	480	507	552	617	700	796	900	1006	1104	1183
3	4	9	15	22	30	40	51	64	80	98	120	146	177	213	254	302	355	412	467	512	542	572	622	693	780	877	977	1070	1146	1194
4	5	10	17	24	33	44	56	71	89	110	134	164	199	240	288	342	402	465	525	574	608	641	695	770	858	950	1038	1109	1155	1170
5	5	11	18	27	36	48	62	79	99	123	151	185	225	271	325	386	453	521	586	638	675	711	768	842	925	1006	1072	1115	1129	1113
6	6	12	20	29	40	53	69	88	111	138	170	209	254	307	368	435	507	579	645	699	738	778	834	903	975	1035	1075	1088	1073	1034
7	6	13	22	32	45	59	77	99	124	155	192	236	287	347	414	486	561	633	698	751	792	832	885	945	998	1034	1046	1032	996	942
8	7	14	24	36	49	66	86	111	140	175	217	267	325	390	462	536	610	679	738	786	827	868	915	960	991	1002	990	956	906	847
9	7	16	27	40	55	74	97	125	158	198	246	301	365	435	508	582	650	709	758	800	840	880	918	946	956	945	915	869	814	754
10	8	18	30	44	62	83	109	141	179	224	278	339	406	478	550	617	674	720	756	789	827	865	893	905	898	872	831	780	724	668
11	9	20	33	49	69	94	123	159	202	253	312	377	446	516	581	636	679	708	731	757	792	826	845	844	825	790	745	694	641	590
12	10	22	37	56	78	106	140	180	229	284	347	413	481	544	597	637	663	677	689	707	738	769	781	772	745	707	662	614	567	522
13	11	24	42	63	88	120	158	203	256	316	380	445	506	558	596	619	629	632	634	647	675	702	709	695	666	628	586	543	501	462
14	12	28	47	71	100	136	178	228	285	346	409	468	518	554	576	585	583	577	574	583	608	632	636	620	591	556	518	480	444	410
15	14	31	53	80	113	153	200	254	312	372	429	478	514	535	543	540	530	520	514	520	542	563	565	549	523	491	458	425	394	365
16	16	35	60	91	128	172	222	278	335	390	437	473	495	503	500	490	476	463	456	461	480	499	500	486	462	434	405	377	351	326
17	18	40	68	102	143	191	243	298	351	397	433	455	464	462	453	439	423	410	403	407	424	441	442	429	409	385	360	336	314	293
18	20	45	77	115	159	209	261	312	357	392	415	426	426	418	405	390	375	363	356	360	375	390	391	380	362	342	321	301	282	264
19	23	51	86	127	174	224	273	317	352	377	389	391	385	374	360	345	331	321	315	319	332	345	346	337	322	305	287	270	254	239
20	26	57	95	139	186	234	277	313	338	352	356	353	344	332	318	305	293	284	279	283	295	306	308	300	287	272	258	243	229	216
21	29	64	104	149	194	237	273	299	316	322	321	315	305	293	281	270	259	252	248	252	262	273	274	268	258	245	233	220	208	197
22	32	69	112	155	197	234	261	280	289	290	287	279	270	259	249	239	230	224	222	225	234	244	245	240	232	221	210	200	190	180
23	35	74	116	157	194	223	244	256	260	259	254	247	238	229	220	212	205	200	198	201	210	218	220	216	209	200	191	182	173	165
24	37	78	118	155	186	208	223	230	232	230	225	218	211	203	196	189	184	180	178	181	189	197	199	195	189	182	174	166	159	152
25	39	78	116	148	173	190	201	205	206	203	199	193	187	181	175	169	165	162	161	164	171	178	180	177	172	166	159	153	146	140
26	39	77	111	138	158	171	179	182	182	180	176	171	166	161	156	152	148	146	146	148	155	161	163	161	157	152	146	140	135	129
27	39	74	103	126	142	152	159	161	161	159	156	152	148	144	141	137	134	132	132	135	141	147	149	147	144	139	134	130	125	120
28	37	68	94	113	127	135	140	142	142	141	139	136	133	130	127	124	122	121	121	123	129	134	136	135	132	128	124	120	116	112
29	34	62	85	101	112	120	124	126	126	125	124	122	120	117	115	113	111	110	110	113	118	123	125	124	122	118	115	111	108	104
30	31	56	76	89	99	106	110	112	112	112	111	110	108	106	104	103	102	101	101	104	108	113	115	114	112	110	107	103	100	97

Substitutes / Nominal, before shock, type y

Average price of other firms

selling price	1	2	3	4	5	6	7	8	9	10	11	12	13	14	15	16	17	18	19	20	21	22	23	24	25	26	27	28	29	30
1	2	4	6	8	11	14	18	21	25	30	34	40	45	49	53	57	62	68	76	85	95	107	121	136	154	175	199	227	259	296
2	2	4	6	9	12	15	19	23	27	32	37	42	48	53	57	61	66	73	82	92	103	116	131	149	169	192	219	251	287	330
3	2	4	7	9	13	16	20	24	29	34	40	45	51	57	61	66	71	79	88	99	112	127	144	163	186	212	243	279	321	370
4	2	4	7	10	13	17	21	26	31	36	42	49	55	61	66	71	77	85	96	108	122	138	157	179	205	235	270	311	359	415
5	2	4	8	11	14	18	23	27	33	39	46	53	60	66	71	77	84	93	104	118	133	152	173	198	227	261	301	348	403	468
6	2	5	8	11	15	19	24	30	36	42	49	57	65	72	77	83	91	101	113	128	146	167	191	219	252	291	337	391	454	529
7	3	5	9	12	16	20	26	32	38	46	53	62	71	78	84	90	99	110	124	141	161	184	212	244	282	326	379	441	513	599
8	3	6	9	13	17	22	28	34	41	49	58	67	77	85	92	99	108	121	136	155	178	204	235	272	315	366	427	498	581	677
9	3	6	10	13	19	24	30	37	45	54	63	74	84	93	101	108	118	132	150	171	197	227	263	305	354	413	482	563	657	764
10	3	6	10	15	20	26	33	40	49	58	69	81	92	103	111	119	130	146	166	190	219	253	294	342	400	467	546	637	741	857
11	3	6	11	16	22	28	35	43	53	64	76	89	102	113	122	131	144	162	184	212	245	284	331	386	452	528	617	719	832	952
12	3	7	12	17	23	30	38	48	58	70	83	98	112	125	135	145	160	180	205	237	274	319	373	436	511	597	696	806	923	1042
13	4	8	13	19	25	33	42	52	64	77	92	108	125	139	150	161	177	200	229	265	309	360	422	494	578	674	781	895	1010	1117
14	4	8	14	20	28	36	46	57	70	85	102	120	139	155	167	180	198	224	258	299	348	407	477	558	652	755	866	978	1083	1168
15	4	9	15	22	30	39	50	63	78	94	113	134	155	173	187	201	222	252	290	337	394	461	539	630	730	838	947	1048	1131	1186
16	5	10	17	24	33	43	55	70	86	104	126	150	174	195	210	226	250	284	327	381	445	521	608	705	810	915	1014	1094	1148	1168
17	5	11	18	26	36	48	61	77	96	117	142	169	196	220	237	255	282	321	370	431	503	587	681	781	884	979	1057	1110	1130	1118
18	6	12	20	29	40	53	68	86	107	132	159	190	221	248	268	288	319	363	419	487	567	657	754	852	944	1020	1072	1092	1081	1043
19	6	13	22	32	44	59	76	96	120	148	180	215	250	281	303	326	361	410	473	548	634	726	821	910	983	1033	1053	1044	1008	953
20	7	14	24	35	49	65	85	107	135	167	203	243	283	317	343	369	408	463	532	612	700	790	875	946	994	1014	1006	973	921	858
21	7	16	27	39	55	73	95	121	153	189	230	275	319	358	387	416	459	520	593	675	760	841	909	955	975	968	937	888	829	765
22	8	18	30	44	61	82	107	137	173	214	260	310	359	402	434	466	514	577	652	731	807	871	916	936	930	901	855	799	739	678
23	9	20	33	49	69	92	121	155	195	242	293	348	401	447	482	517	567	632	703	774	834	876	896	891	864	822	769	712	655	599
24	10	22	37	55	77	104	137	175	221	272	328	386	442	490	527	565	616	677	741	796	836	855	851	827	788	739	686	631	579	530
25	11	24	41	62	87	118	155	198	248	304	363	422	478	526	565	605	654	708	758	795	814	811	790	754	708	658	607	558	512	469
26	12	28	47	70	99	134	175	223	277	335	395	452	505	550	591	631	676	720	753	771	770	751	719	677	631	583	537	493	453	416
27	14	31	53	79	112	151	196	248	305	364	421	473	519	560	600	640	678	709	727	727	712	683	645	603	559	516	475	437	402	370
28	16	35	60	89	126	169	219	273	330	386	436	480	517	552	591	629	660	679	682	671	646	613	574	534	494	456	420	387	358	331
29	18	40	67	101	142	188	240	295	349	398	439	472	500	530	565	601	625	633	627	608	579	545	509	472	437	403	373	345	320	297
30	20	45	76	113	158	206	259	311	358	398	429	452	471	495	527	559	578	579	566	544	515	483	450	417	387	358	332	308	287	267

Substitutes / Nominal, after shock, type *x*

Average price of other firms

selling price	1	2	3	4	5	6	7	8	9	10	11	12	13	14	15	16	17	18	19	20	21	22	23	24	25	26	27	28	29	30
1	5	13	27	49	85	137	181	231	326	398	382	306	230	173	133	105	85	71	60	52	45	40	35	32	29	26	24	22	21	19
2	7	18	37	69	123	193	246	301	358	344	279	212	161	124	99	81	67	57	49	43	38	34	31	28	26	23	22	20	19	18
3	9	24	53	100	169	236	280	315	305	251	193	148	116	93	76	64	54	47	41	37	33	30	27	25	23	21	19	18	17	16
4	12	35	76	137	199	226	246	257	221	174	135	107	86	71	60	51	45	39	35	31	28	26	24	22	20	19	18	17	16	14
5	18	51	104	159	181	173	181	183	153	122	98	80	66	56	48	42	37	33	30	27	25	23	21	20	18	17	16	15	14	13
6	26	70	119	142	135	121	125	127	107	88	73	61	52	45	40	35	32	29	26	24	22	20	19	18	16	15	15	13	13	12
7	35	79	105	105	94	84	87	89	78	66	56	48	42	37	33	30	27	25	23	21	19	18	17	16	15	14	13	13	12	12
8	39	69	77	73	65	60	63	65	58	51	44	39	35	31	28	26	24	22	20	19	17	16	15	14	14	12	12	12	11	11
9	34	50	54	51	47	44	47	49	45	40	36	32	29	26	24	22	21	19	18	17	16	15	14	13	12	12	11	11	10	10
10	25	35	37	37	35	34	36	38	36	32	29	27	25	23	21	19	18	17	16	15	14	13	13	12	11	11	10	10	10	9
11	17	24	27	27	27	26	29	30	29	27	25	23	21	20	18	17	16	15	14	14	13	12	12	11	11	10	10	9	9	9
12	12	18	20	21	21	21	23	25	24	22	21	20	18	17	16	15	14	14	13	12	12	11	11	10	10	9	9	9	8	8
13	9	13	15	16	17	17	19	20	20	19	18	17	16	15	14	14	13	12	12	11	11	10	10	9	9	9	8	8	8	8
14	7	10	12	13	14	14	16	17	17	16	16	15	14	13	13	12	12	11	11	10	10	9	9	9	8	8	8	8	7	7
15	5	8	10	11	12	12	13	15	15	14	14	13	13	12	12	11	11	10	10	9	9	9	8	8	8	8	7	7	7	7
16	4	6	8	9	10	10	12	13	13	12	12	12	11	11	10	10	10	9	9	9	8	8	8	8	7	7	7	7	7	6
17	3	5	7	8	8	8	10	11	11	11	11	10	10	10	9	9	9	8	8	8	8	8	7	7	7	7	7	6	6	6
18	3	4	6	7	7	8	9	10	10	10	10	9	9	9	9	8	8	8	8	7	7	7	7	7	6	6	6	6	6	6
19	2	4	5	5	6	7	8	8	9	9	9	8	8	8	8	7	8	7	7	7	7	7	6	6	6	6	6	6	6	5
20	2	3	4	5	6	6	7	8	8	8	8	8	8	7	7	7	7	7	7	6	6	6	6	6	6	6	5	5	5	5
21	2	2	4	4	5	5	6	7	7	7	7	7	7	6	7	7	6	6	6	6	6	6	6	6	5	5	5	5	5	5
22	1	2	3	4	4	5	6	6	6	6	6	6	6	6	6	6	6	6	6	6	6	5	5	5	5	5	5	5	5	5
23	1	2	3	4	4	4	5	6	6	6	6	5	6	6	6	6	6	6	5	5	5	5	5	5	5	5	5	5	5	4
24	1	2	3	3	4	4	5	5	5	5	5	5	5	5	5	5	5	5	5	5	5	5	5	5	5	5	4	4	4	4
25	1	2	2	3	3	4	4	5	5	4	5	5	5	5	5	5	5	5	5	5	5	5	5	4	4	4	4	4	4	4
26	1	2	2	3	3	3	4	4	4	4	5	5	5	5	5	5	5	5	5	4	4	4	4	4	4	4	4	4	4	4
27	1	1	2	2	3	3	3	3	4	4	4	4	4	4	4	4	4	4	4	4	4	4	4	4	4	4	4	4	4	4
28	1	1	2	2	3	2	3	3	4	4	4	4	4	4	4	4	4	4	4	4	4	4	4	4	4	4	4	4	4	4
29	1	1	2	1	2	2	3	3	4	4	4	4	4	4	4	4	4	4	4	4	4	4	4	4	4	4	4	3	3	3
30	1	1	1	1	2	2	3	3	3	3	3	4	4	4	4	4	4	4	4	4	4	4	4	3	3	3	3	3	3	3

Substitutes / Nominal, after shock, type *y*

Average price of other firms

selling price	1	2	3	4	5	6	7	8	9	10	11	12	13	14	15	16	17	18	19	20	21	22	23	24	25	26	27	28	29	30
1	2	5	10	15	20	26	37	54	81	123	192	300	440	541	513	402	294	217	164	127	102	84	70	60	52	46	41	36	33	30
2	3	6	12	19	26	34	49	73	112	176	277	410	506	481	378	278	206	156	122	98	81	68	58	50	44	39	35	32	29	26
3	3	8	15	25	34	44	66	102	161	254	379	470	449	355	262	194	148	116	93	77	65	56	48	43	38	34	31	28	26	24
4	4	10	19	33	45	60	91	145	232	347	433	416	331	246	183	140	110	89	74	62	53	46	41	36	33	30	27	25	23	21
5	5	13	26	45	62	84	131	210	316	395	382	305	229	172	132	104	84	70	59	51	45	39	35	32	29	26	24	22	21	19
6	7	18	36	63	89	121	189	284	357	348	281	212	160	124	98	80	67	57	49	43	38	34	30	28	25	23	21	20	19	17
7	9	24	51	92	129	173	253	318	312	255	195	148	115	92	75	63	54	46	41	36	32	29	27	24	22	21	19	18	17	16
8	12	34	74	129	176	226	279	275	229	177	136	107	86	71	59	51	44	39	35	31	28	26	23	22	20	19	17	16	15	14
9	18	50	102	157	200	236	237	201	158	123	98	79	66	56	48	42	37	33	30	27	25	23	21	19	18	17	16	15	14	13
10	25	68	119	151	176	193	172	139	111	89	73	61	52	45	39	35	31	28	26	24	22	20	19	17	16	15	14	13	13	12
11	35	80	109	115	129	137	119	97	80	66	56	48	42	37	33	30	27	24	22	21	19	18	17	16	15	14	13	13	12	11
12	40	71	81	81	89	95	83	70	60	51	44	39	34	31	28	25	23	21	20	18	17	16	15	14	14	12	12	12	11	11
13	35	52	56	56	62	67	60	53	46	40	36	32	29	26	24	22	20	19	18	16	15	15	14	13	12	12	11	11	10	10
14	26	36	39	40	45	49	45	40	36	33	29	27	24	22	21	19	18	17	16	15	14	13	13	11	11	11	10	10	10	9
15	18	25	28	29	34	37	35	32	29	27	25	23	21	19	18	17	16	15	14	13	13	12	11	11	10	10	10	9	9	9
16	12	18	21	22	26	29	28	26	24	22	21	19	18	17	16	15	14	13	13	12	12	11	11	10	10	9	9	9	8	8
17	9	14	16	17	20	23	22	21	20	19	18	17	16	15	14	13	13	12	12	11	11	10	10	9	9	9	8	8	8	8
18	7	10	13	14	16	18	18	18	17	16	15	15	14	13	13	12	12	11	11	10	10	9	9	9	8	8	8	8	7	7
19	5	8	10	12	14	15	15	15	15	14	14	13	12	12	11	11	11	10	10	9	9	9	8	8	8	8	7	7	7	7
20	4	6	8	9	11	13	13	13	13	12	12	12	11	11	10	10	10	9	9	9	8	8	8	8	7	7	7	7	6	6
21	3	5	7	8	10	11	11	11	11	11	11	10	10	10	9	9	9	8	8	8	8	7	7	7	7	7	6	6	6	6
22	3	4	6	7	8	9	10	10	10	10	10	9	9	9	9	8	8	8	8	7	7	7	7	7	6	6	6	6	6	6
23	2	4	5	5	7	8	9	9	9	9	9	8	8	8	8	7	7	7	7	7	7	7	6	6	6	6	6	6	5	5
24	2	3	4	5	6	7	8	8	8	8	8	8	7	7	7	7	7	7	7	6	6	6	6	6	6	6	5	5	5	5
25	2	2	4	5	6	6	7	7	7	7	7	7	7	6	7	7	6	6	6	6	6	6	6	5	5	5	5	5	5	5
26	1	2	3	4	5	6	6	6	6	6	6	6	6	6	6	6	6	6	6	6	6	5	5	5	5	5	5	5	5	5
27	1	2	3	4	4	5	5	6	6	6	6	5	6	6	6	6	6	5	5	5	5	5	5	5	5	5	5	5	4	4
28	1	2	3	3	4	5	5	5	5	5	5	5	5	5	5	5	5	5	5	5	5	5	5	5	5	4	4	4	4	4
29	1	2	2	3	4	4	4	5	5	4	5	5	5	5	5	5	5	5	5	5	5	5	4	4	4	4	4	4	4	4
30	1	2	2	3	3	4	4	4	4	4	5	5	5	5	5	5	5	5	4	4	4	4	4	4	4	4	4	4	4	4

Substitutes / Pure Real, before shock, type *x*

Average price of other firms

selling price	3	6	9	12	15	18	21	24	27	30	33	36	39	42	45	48	51	54	57	60	63	66	69	72	75	78	81	84	87	90
3	5	7	9	12	17	23	26	29	36	40	35	26	18	12	9	7	5	4	3	3	2	2	2	1	1	1	1	1	1	1
6	7	9	12	17	25	32	35	38	40	34	25	18	12	9	7	5	4	3	3	2	2	2	1	1	1	1	1	1	1	1
9	9	12	18	25	34	39	40	39	34	25	18	12	9	7	5	4	3	3	2	2	2	1	1	1	1	1	1	1	1	1
12	12	18	25	34	40	38	35	32	25	17	12	9	7	5	4	3	3	2	2	2	1	1	1	1	1	1	1	1	1	0
15	18	26	35	40	36	29	26	23	17	12	9	7	5	4	3	3	2	2	2	1	1	1	1	1	1	1	1	1	0	0
18	26	35	40	36	27	20	18	16	12	9	7	5	4	3	3	2	2	2	1	1	1	1	1	1	1	1	1	0	0	0
21	35	40	35	26	19	14	12	11	9	7	5	4	3	3	2	2	2	1	1	1	1	1	1	1	1	1	0	0	0	0
24	39	35	26	18	13	10	9	8	6	5	4	3	3	2	2	2	1	1	1	1	1	1	1	1	1	0	0	0	0	0
27	34	25	18	13	9	7	7	6	5	4	3	3	2	2	2	1	1	1	1	1	1	1	1	1	0	0	0	0	0	0
30	25	18	12	9	7	6	5	5	4	3	3	2	2	2	1	1	1	1	1	1	1	1	1	1	0	0	0	0	0	0
33	17	12	9	7	5	4	4	4	3	3	2	2	2	1	1	1	1	1	1	1	1	1	1	0	0	0	0	0	0	0
36	12	9	7	5	4	4	3	3	3	2	2	2	1	1	1	1	1	1	1	1	1	1	0	0	0	0	0	0	0	0
39	9	7	5	4	3	3	3	3	2	2	2	1	1	1	1	1	1	1	1	1	1	0	0	0	0	0	0	0	0	0
42	7	5	4	3	3	2	2	2	2	2	1	1	1	1	1	1	1	1	1	1	0	0	0	0	0	0	0	0	0	0
45	5	4	3	3	2	2	2	2	2	1	1	1	1	1	1	1	1	1	1	0	0	0	0	0	0	0	0	0	0	0
48	4	3	3	2	2	2	2	2	1	1	1	1	1	1	1	1	1	1	0	0	0	0	0	0	0	0	0	0	0	0
51	3	3	2	2	2	1	1	1	1	1	1	1	1	1	1	1	1	0	0	0	0	0	0	0	0	0	0	0	0	0
54	3	2	2	2	1	1	1	1	1	1	1	1	1	1	1	1	0	0	0	0	0	0	0	0	0	0	0	0	0	0
57	2	2	2	1	1	1	1	1	1	1	1	1	1	1	1	0	0	0	0	0	0	0	0	0	0	0	0	0	0	0
60	2	2	1	1	1	1	1	1	1	1	1	1	1	1	0	0	0	0	0	0	0	0	0	0	0	0	0	0	0	0
63	2	1	1	1	1	1	1	1	1	1	1	1	1	0	0	0	0	0	0	0	0	0	0	0	0	0	0	0	0	0
66	1	1	1	1	1	1	1	1	1	1	1	1	0	0	0	0	0	0	0	0	0	0	0	0	0	0	0	0	0	0
69	1	1	1	1	1	1	1	1	1	1	1	0	0	0	0	0	0	0	0	0	0	0	0	0	0	0	0	0	0	0
72	1	1	1	1	1	1	1	1	1	1	0	0	0	0	0	0	0	0	0	0	0	0	0	0	0	0	0	0	0	0
75	1	1	1	1	1	1	1	1	1	0	0	0	0	0	0	0	0	0	0	0	0	0	0	0	0	0	0	0	0	0
78	1	1	1	1	1	1	1	1	0	0	0	0	0	0	0	0	0	0	0	0	0	0	0	0	0	0	0	0	0	0
81	1	1	1	1	1	1	0	0	0	0	0	0	0	0	0	0	0	0	0	0	0	0	0	0	0	0	0	0	0	0
84	1	1	1	1	1	0	0	0	0	0	0	0	0	0	0	0	0	0	0	0	0	0	0	0	0	0	0	0	0	0
87	1	1	1	0	0	0	0	0	0	0	0	0	0	0	0	0	0	0	0	0	0	0	0	0	0	0	0	0	0	0
90	1	1	0	0	0	0	0	0	0	0	0	0	0	0	0	0	0	0	0	0	0	0	0	0	0	0	0	0	0	0

Substitutes / Pure Real, before shock, type *y*

Average price of other firms

selling price	3	6	9	12	15	18	21	24	27	30	33	36	39	42	45	48	51	54	57	60	63	66	69	72	75	78	81	84	87	90
3	2	3	3	4	4	4	5	7	9	12	17	25	34	39	34	25	17	12	9	6	5	4	3	3	2	2	2	1	1	1
6	3	3	4	5	5	6	7	9	12	18	25	34	39	34	25	17	12	9	6	5	4	3	3	2	2	2	1	1	1	1
9	3	4	5	6	7	7	9	13	18	25	34	39	35	25	17	12	9	6	5	4	3	3	2	2	2	1	1	1	1	1
12	4	5	6	8	9	10	13	18	26	35	39	35	25	18	12	9	6	5	4	3	3	2	2	2	1	1	1	1	1	1
15	5	7	9	11	12	14	19	26	35	40	35	25	18	12	9	7	5	4	3	3	2	2	2	1	1	1	1	1	1	1
18	7	9	12	16	18	20	27	36	40	35	26	18	12	9	7	5	4	3	3	2	2	2	1	1	1	1	1	1	1	1
21	9	12	17	23	26	29	36	40	35	26	18	12	9	7	5	4	3	3	2	2	2	1	1	1	1	1	1	1	1	1
24	12	17	25	32	35	38	40	34	25	18	12	9	7	5	4	3	3	2	2	2	1	1	1	1	1	1	1	1	1	0
27	18	25	34	39	40	39	34	25	18	12	9	7	5	4	3	3	2	2	2	1	1	1	1	1	1	1	1	1	0	0
30	25	34	40	38	35	32	25	17	12	9	7	5	4	3	3	2	2	2	1	1	1	1	1	1	1	1	1	0	0	0
33	35	40	36	29	26	23	17	12	9	7	5	4	3	3	2	2	2	1	1	1	1	1	1	1	1	1	0	0	0	0
36	40	36	27	20	18	16	12	9	7	5	4	3	3	2	2	2	1	1	1	1	1	1	1	1	1	0	0	0	0	0
39	35	26	19	14	12	11	9	7	5	4	3	3	2	2	2	1	1	1	1	1	1	1	1	1	0	0	0	0	0	0
42	26	18	13	10	9	8	6	5	4	3	3	2	2	2	1	1	1	1	1	1	1	1	1	0	0	0	0	0	0	0
45	18	13	9	7	7	6	5	4	3	3	2	2	2	1	1	1	1	1	1	1	1	1	0	0	0	0	0	0	0	0
48	12	9	7	6	5	5	4	3	3	2	2	2	1	1	1	1	1	1	1	1	1	1	0	0	0	0	0	0	0	0
51	9	7	5	4	4	4	3	3	2	2	2	1	1	1	1	1	1	1	1	1	1	0	0	0	0	0	0	0	0	0
54	7	5	4	4	3	3	3	2	2	2	1	1	1	1	1	1	1	1	1	1	0	0	0	0	0	0	0	0	0	0
57	5	4	3	3	3	3	2	2	2	1	1	1	1	1	1	1	1	1	1	0	0	0	0	0	0	0	0	0	0	0
60	4	3	3	2	2	2	2	2	1	1	1	1	1	1	1	1	1	1	0	0	0	0	0	0	0	0	0	0	0	0
63	3	3	2	2	2	2	2	1	1	1	1	1	1	1	1	1	1	0	0	0	0	0	0	0	0	0	0	0	0	0
66	3	2	2	2	2	2	1	1	1	1	1	1	1	1	1	1	0	0	0	0	0	0	0	0	0	0	0	0	0	0
69	2	2	2	1	1	1	1	1	1	1	1	1	1	1	1	0	0	0	0	0	0	0	0	0	0	0	0	0	0	0
72	2	2	1	1	1	1	1	1	1	1	1	1	1	1	0	0	0	0	0	0	0	0	0	0	0	0	0	0	0	0
75	2	1	1	1	1	1	1	1	1	1	1	1	1	0	0	0	0	0	0	0	0	0	0	0	0	0	0	0	0	0
78	1	1	1	1	1	1	1	1	1	1	1	1	0	0	0	0	0	0	0	0	0	0	0	0	0	0	0	0	0	0
81	1	1	1	1	1	1	1	1	1	1	1	0	0	0	0	0	0	0	0	0	0	0	0	0	0	0	0	0	0	0
84	1	1	1	1	1	1	1	1	1	1	0	0	0	0	0	0	0	0	0	0	0	0	0	0	0	0	0	0	0	0
87	1	1	1	1	1	1	1	1	1	0	0	0	0	0	0	0	0	0	0	0	0	0	0	0	0	0	0	0	0	0
90	1	1	1	1	1	1	1	1	0	0	0	0	0	0	0	0	0	0	0	0	0	0	0	0	0	0	0	0	0	0

Substitutes / Pure Real, after shock, type *x*

Average price of other firms

selling price	1	2	3	4	5	6	7	8	9	10	11	12	13	14	15	16	17	18	19	20	21	22	23	24	25	26	27	28	29	30
1	5	7	9	12	17	23	26	29	36	40	35	26	18	12	9	7	5	4	3	3	2	2	2	1	1	1	1	1	1	1
2	7	9	12	17	25	32	35	38	40	34	25	18	12	9	7	5	4	3	3	2	2	2	1	1	1	1	1	1	1	1
3	9	12	18	25	34	39	40	39	34	25	18	12	9	7	5	4	3	3	2	2	2	1	1	1	1	1	1	1	1	1
4	12	18	25	34	40	38	35	32	25	17	12	9	7	5	4	3	3	2	2	2	1	1	1	1	1	1	1	1	1	0
5	18	26	35	40	36	29	26	23	17	12	9	7	5	4	3	3	2	2	2	1	1	1	1	1	1	1	1	1	0	0
6	26	35	40	36	27	20	18	16	12	9	7	5	4	3	3	2	2	2	1	1	1	1	1	1	1	1	1	0	0	0
7	35	40	35	26	19	14	12	11	9	7	5	4	3	3	2	2	2	1	1	1	1	1	1	1	1	1	0	0	0	0
8	39	35	26	18	13	10	9	8	6	5	4	3	3	2	2	2	1	1	1	1	1	1	1	1	1	0	0	0	0	0
9	34	25	18	13	9	7	7	6	5	4	3	3	2	2	2	1	1	1	1	1	1	1	1	1	0	0	0	0	0	0
10	25	18	12	9	7	6	5	5	4	3	3	2	2	2	1	1	1	1	1	1	1	1	1	1	0	0	0	0	0	0
11	17	12	9	7	5	4	4	4	3	3	2	2	2	1	1	1	1	1	1	1	1	1	1	0	0	0	0	0	0	0
12	12	9	7	5	4	4	3	3	3	2	2	2	1	1	1	1	1	1	1	1	1	1	0	0	0	0	0	0	0	0
13	9	7	5	4	3	3	3	3	2	2	2	1	1	1	1	1	1	1	1	1	1	0	0	0	0	0	0	0	0	0
14	7	5	4	3	3	2	2	2	2	2	1	1	1	1	1	1	1	1	1	1	0	0	0	0	0	0	0	0	0	0
15	5	4	3	3	2	2	2	2	2	1	1	1	1	1	1	1	1	1	1	0	0	0	0	0	0	0	0	0	0	0
16	4	3	3	2	2	2	2	2	1	1	1	1	1	1	1	1	1	1	0	0	0	0	0	0	0	0	0	0	0	0
17	3	3	2	2	2	1	1	1	1	1	1	1	1	1	1	1	1	0	0	0	0	0	0	0	0	0	0	0	0	0
18	3	2	2	2	1	1	1	1	1	1	1	1	1	1	1	1	0	0	0	0	0	0	0	0	0	0	0	0	0	0
19	2	2	2	1	1	1	1	1	1	1	1	1	1	1	1	0	0	0	0	0	0	0	0	0	0	0	0	0	0	0
20	2	2	1	1	1	1	1	1	1	1	1	1	1	1	0	0	0	0	0	0	0	0	0	0	0	0	0	0	0	0
21	2	1	1	1	1	1	1	1	1	1	1	1	1	0	0	0	0	0	0	0	0	0	0	0	0	0	0	0	0	0
22	1	1	1	1	1	1	1	1	1	1	1	1	0	0	0	0	0	0	0	0	0	0	0	0	0	0	0	0	0	0
23	1	1	1	1	1	1	1	1	1	1	1	0	0	0	0	0	0	0	0	0	0	0	0	0	0	0	0	0	0	0
24	1	1	1	1	1	1	1	1	1	1	0	0	0	0	0	0	0	0	0	0	0	0	0	0	0	0	0	0	0	0
25	1	1	1	1	1	1	1	1	1	0	0	0	0	0	0	0	0	0	0	0	0	0	0	0	0	0	0	0	0	0
26	1	1	1	1	1	1	1	1	0	0	0	0	0	0	0	0	0	0	0	0	0	0	0	0	0	0	0	0	0	0
27	1	1	1	1	1	1	0	0	0	0	0	0	0	0	0	0	0	0	0	0	0	0	0	0	0	0	0	0	0	0
28	1	1	1	1	1	0	0	0	0	0	0	0	0	0	0	0	0	0	0	0	0	0	0	0	0	0	0	0	0	0
29	1	1	1	0	0	0	0	0	0	0	0	0	0	0	0	0	0	0	0	0	0	0	0	0	0	0	0	0	0	0
30	1	1	0	0	0	0	0	0	0	0	0	0	0	0	0	0	0	0	0	0	0	0	0	0	0	0	0	0	0	0

Substitutes / Pure Real, after shock, type *y*

Average price of other firms

selling price	1	2	3	4	5	6	7	8	9	10	11	12	13	14	15	16	17	18	19	20	21	22	23	24	25	26	27	28	29	30
1	2	3	3	4	4	4	5	7	9	12	17	25	34	39	34	25	17	12	9	6	5	4	3	3	2	2	2	1	1	1
2	3	3	4	5	5	6	7	9	12	18	25	34	39	34	25	17	12	9	6	5	4	3	3	2	2	2	1	1	1	1
3	3	4	5	6	7	7	9	13	18	25	34	39	35	25	17	12	9	6	5	4	3	3	2	2	2	1	1	1	1	1
4	4	5	6	8	9	10	13	18	26	35	39	35	25	18	12	9	6	5	4	3	3	2	2	2	1	1	1	1	1	1
5	5	7	9	11	12	14	19	26	35	40	35	25	18	12	9	7	5	4	3	3	2	2	2	1	1	1	1	1	1	1
6	7	9	12	16	18	20	27	36	40	35	26	18	12	9	7	5	4	3	3	2	2	2	1	1	1	1	1	1	1	1
7	9	12	17	23	26	29	36	40	35	26	18	12	9	7	5	4	3	3	2	2	2	1	1	1	1	1	1	1	1	1
8	12	17	25	32	35	38	40	34	25	18	12	9	7	5	4	3	3	2	2	2	1	1	1	1	1	1	1	1	1	0
9	18	25	34	39	40	39	34	25	18	12	9	7	5	4	3	3	2	2	2	1	1	1	1	1	1	1	1	1	0	0
10	25	34	40	38	35	32	25	17	12	9	7	5	4	3	3	2	2	2	1	1	1	1	1	1	1	1	1	0	0	0
11	35	40	36	29	26	23	17	12	9	7	5	4	3	3	2	2	2	1	1	1	1	1	1	1	1	1	0	0	0	0
12	40	36	27	20	18	16	12	9	7	5	4	3	3	2	2	2	1	1	1	1	1	1	1	1	1	0	0	0	0	0
13	35	26	19	14	12	11	9	7	5	4	3	3	2	2	2	1	1	1	1	1	1	1	1	1	0	0	0	0	0	0
14	26	18	13	10	9	8	6	5	4	3	3	2	2	2	1	1	1	1	1	1	1	1	1	0	0	0	0	0	0	0
15	18	13	9	7	7	6	5	4	3	3	2	2	2	1	1	1	1	1	1	1	1	1	0	0	0	0	0	0	0	0
16	12	9	7	6	5	5	4	3	3	2	2	2	1	1	1	1	1	1	1	1	1	1	0	0	0	0	0	0	0	0
17	9	7	5	4	4	4	3	3	2	2	2	1	1	1	1	1	1	1	1	1	1	0	0	0	0	0	0	0	0	0
18	7	5	4	4	3	3	3	2	2	2	1	1	1	1	1	1	1	1	1	1	0	0	0	0	0	0	0	0	0	0
19	5	4	3	3	3	3	2	2	2	1	1	1	1	1	1	1	1	1	1	0	0	0	0	0	0	0	0	0	0	0
20	4	3	3	2	2	2	2	2	1	1	1	1	1	1	1	1	1	1	0	0	0	0	0	0	0	0	0	0	0	0
21	3	3	2	2	2	2	2	1	1	1	1	1	1	1	1	1	1	0	0	0	0	0	0	0	0	0	0	0	0	0
22	3	2	2	2	2	2	1	1	1	1	1	1	1	1	1	1	0	0	0	0	0	0	0	0	0	0	0	0	0	0
23	2	2	2	1	1	1	1	1	1	1	1	1	1	1	1	0	0	0	0	0	0	0	0	0	0	0	0	0	0	0
24	2	2	1	1	1	1	1	1	1	1	1	1	1	1	0	0	0	0	0	0	0	0	0	0	0	0	0	0	0	0
25	2	1	1	1	1	1	1	1	1	1	1	1	1	0	0	0	0	0	0	0	0	0	0	0	0	0	0	0	0	0
26	1	1	1	1	1	1	1	1	1	1	1	1	0	0	0	0	0	0	0	0	0	0	0	0	0	0	0	0	0	0
27	1	1	1	1	1	1	1	1	1	1	1	0	0	0	0	0	0	0	0	0	0	0	0	0	0	0	0	0	0	0
28	1	1	1	1	1	1	1	1	1	1	0	0	0	0	0	0	0	0	0	0	0	0	0	0	0	0	0	0	0	0
29	1	1	1	1	1	1	1	1	1	0	0	0	0	0	0	0	0	0	0	0	0	0	0	0	0	0	0	0	0	0
30	1	1	1	1	1	1	1	1	0	0	0	0	0	0	0	0	0	0	0	0	0	0	0	0	0	0	0	0	0	0

A3. Mathematical appendix

Section 3.1. discusses the functional form which was used to generate payoff tables, section 3.2. illustrates the loss decomposition, and 3.3. explains techniques how subjects can find the equilibrium.

3.1. Payoff function

The real payoff π^t_{ik} of each agent *i* of type *k* in every period $t \in [0, 40]$ depends on the price P_i he chooses and on the average price all other *-i* agents choose. Price choices have to be from $P^t_{ik} \in [1, 30]$. The average price for agent *i* ($\bar{P}^t_{-ik}$) is the simple average of all prices chosen by *other* agents in period *t*. Specifically, we used groups of $n = 4$ agents, with 2 agents of type *x* and *y*, respectively. For example, the average price for agent 1 of type *x* is: $\bar{P}_{1x} = (P_{2x} + P_{1y} + P_{2y})/3$.
The real payoff for agent *i* of type *k* in period *t* was calculated by:

(1)

$$\pi^t_{ik} = \frac{V \cdot \left[\frac{1 + a\left(\frac{\bar{P}^t_{-ik}}{M_j} - \frac{\bar{P}^*_{kj}}{M_j}\right)^2}{1 + b\left(\frac{\bar{P}^t_{-ik}}{M_j} - \frac{\bar{P}^*_{kj}}{M_j}\right)^2}\right]}{1 + c\left\{\left(\frac{P^t_{ik}}{M_j} - \frac{P^*_{kj}}{M_j}\right) - d\left(\frac{\bar{P}^t_{-ik}}{M_j} - \frac{\bar{P}^*_{kj}}{M_j}\right) + e \cdot \arctan\left[\left(\frac{\bar{P}^t_{-ik}}{M_j} - \frac{\bar{P}^*_{kj}}{M_j}\right) \cdot f\right]\right\}^2}$$

where: P^*_{kj}, $\bar{P}^*_{kj}$ is the equilibrium nominal price and the equilibrium nominal average price, respectively, for an agent of type *k* in the pre-shock ($j = 0$ in $t \in [0, 20]$) or post-shock phase ($j = 1$ in $t \in [21, 40]$). M_j is the quantity of money in the pre- (M_0) or post-shock phase. *a, b, c, d, e, f, V,* are constants in all periods. The only difference between strategic complements treatment (CT) and strategic substitutes (ST) is that *d* and *e* are change signs. The only difference between types of agents *k* is the level of equilibrium prices P^*_{kj}, $\bar{P}^*_{kj}$. These change equiproportionately to the quantity of money (e.g. $P^*_{k0}/M_0 = P^*_{k1}/M_1$). *Table* 16 provides the numerical values of all parameters.

Table 16: Parametrization of the payoff function

Constant (all periods)	Pre-shock	Post-shock
$a = 0.5$	$P^*_{x0} = 9$	$P^*_{y0} = 27$
$b = 0.6$	$P^*_{x1} = 3$	$P^*_{y1} = 9$
$c = 27$	$\bar{P}^*_{x0} = 21$	$\bar{P}^*_{y0} = 15$
$d = \pm 1$	$\bar{P}^*_{x1} = 7$	$\bar{P}^*_{y1} = 5$
$e = \pm 0.05$	$M_0 = 42$	
$f = 20$	$M_1 = 14$	
$V = 40$		

Note: $d, e > 0$ for strategic complements;
$d, e < 0$ for strategic substitutes.

A) Properties of the payoff function

The payoff function obviously is *homogeneous of degree zero* in the quantity of money (M), own chosen prices (P_i) and average prices (P_{-i}). That is

$\lambda^r \cdot \pi_i(P_i, P_{-i}, M) = \pi_i(\lambda P_i, \lambda P_{-i}, \lambda M)$ with $\lambda > 0$, and $r = 0$ for all $i = 1, \ldots, n$.

Suppressing all indices for types, periods, subjects and shock conditions, the payoff function is more conveniently written as:

$$\text{(2)} \qquad \pi(p, \bar{p}) = \frac{VA(\bar{p})}{1 + cB^2(p, \bar{p})},$$

with $p = p_i = P_i/M$, $\bar{p} = \bar{p}_{-i} = P_{-i}/M$, $A(\bar{p}) = \dfrac{1 + a(\bar{p} - \bar{p}^*)^2}{1 + b(\bar{p} - \bar{p}^*)^2}$, and

$B(p, \bar{p}) = (p - p^*) - d(\bar{p} - \bar{p}^*) + e \cdot \arctan[(\bar{p} - \bar{p}^*) \cdot f]$

The payoff function is symmetric and quasiconcave in p, having a bell-shaped curvature. This can be seen by noting that the payoff function is of the type $y(p) = 1/(1 + p^2)$. The payoff function takes the maximum value of $\pi^* = V$ for all i, k, t if all prices are in equilibrium, i.e. $p = p^*$. Hence, the equilibrium is pareto-efficient. Since the payoff function is continously differentiable, the necessary first order condition for a maximum is:

(3) $$\frac{\partial \pi(p, \bar{p})}{\partial p} = -\frac{2cVABB_p}{(1 + cB^2)^2} = 0.$$

Which implies that either $A(\bar{p}) = 0$, $B(p, \bar{p}) = 0$, or $B_p(p, \bar{p}) = 0$.

But $A \geq 0$, since $a > 0$, and $B_p = 1$. Thus, we must have $B = 0$, i.e.

$(p - p^*) - d(\bar{p} - \bar{p}^*) + e \cdot \arctan[(\bar{p} - \bar{p}^*) \cdot f] = 0$.

With the deviation of real equilibrium prices of their equilibrium values denoted as $D = (\bar{p} - \bar{p}^*)$ this is more conveniently written as

(4) $$p_i^{br}(\bar{p}) = dD - (e \cdot \arctan(Df)) + p^*$$

which is the *best reply function* for all i.

Equation (4) shows that it is a best reply to choose the equilibrium p^* if all other players choose equilibrium prices $\bar{p}^*$ (in this case, $D = 0$). If other players choose prices above the equilibrium (i.e. $\bar{p} > \bar{p}^*$ and $D > 0$) it is a best reply to choose above equilibrium prices ($p > p^*$) if strategic complements prevail ($d > 0$), but to choose prices below the equilibrium if strategic substitutes prevail ($d > 0$).

Differentiating the best reply function with respect to the average price yields

(5) $$\left.\frac{\partial p^{br}}{\partial \bar{p}}\right|_{\frac{\partial \pi}{\partial p} = 0} = d - \frac{e}{1 + D^2 f}.$$

If $d > e > 0$, the slope of the best reply function is positive (since $f > 0$). If $d < e < 0$, the slope is negative. In the former case, we have strategic complementarity, in the latter strategic substitutability. Note that the slope is close to d in the CT if other players choose disequilibrium prices since the second term in (5) approaches zero for D large. Hence, the slope of the best reply away from equilibrium is -1 in the ST, and +1 in the CT. Note that this is the maximum degree of complementarity that can be implemented without obtaining multiple equilibria. The arcus tangens function in (4) causes the best reply function to be S-shaped. This feature has been introduced to make equilibrium play less risky and more stable. The reason for increased stability is that if the other -i players (for whatever reason) slightly deviate from the equilibrium choice, it is still a best reply to choose the equilibrium for player i.

3.2. Loss decomposition

Since maximum possible profits prevail in equilibrium ($\pi^* = \pi_i(p_i^*, \bar{p}_{-i}^*) = V$) for all i and t, individual losses occur whenever choices are in disequilibrium. In principle, losses can occur for three reasons: First, a subject may not choose a best reply to her correct expectation. Second, although a subject may play best reply given her expectation, her expectations may be wrong. Third, a subject may incur a loss despite behaving rationally given her expectation and despite holding the correct expectation if other subjects choose disequilibrium prices. The loss decomposition is a technique to investigate the relative importance of these losses.

Figure 40: Illustration of loss decomposition

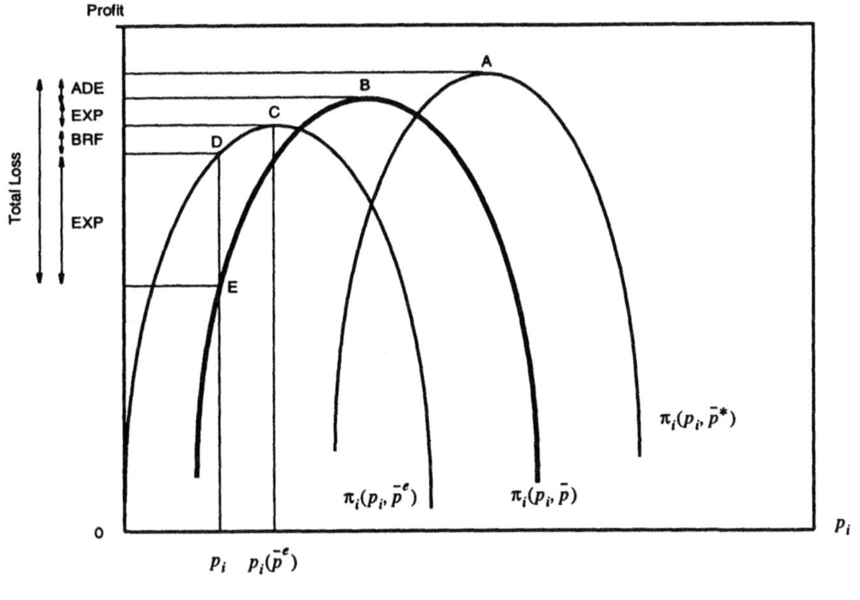

Let $p_i = P_i/M$ be the real price that agent i (for simplicity, we suppress the indices for types k and periods t). Let $\bar{p}_{-i}$ be the actual average real price of the other n -1 players in the group, and let $\bar{p}^e_{-i}$ be the average real price that agent i expected to materialize in this period. Let $p_i(\bar{p}_{-i})$ be agent i's best reply on the actual average price, and $p_i(\bar{p}^e_{-i})$ agent i's best reply on the expected average price. In addition, p^*_i and $\bar{p}^*_{-i}$ denote equilibrium values for agent i. Then, we define:

$\pi_i^* = \pi_i(p_i^*, \bar{p}_{-i}^*)$	Equilibrium payoff (point A in *figure* 40)
$\pi_i^{Hyp} = \pi_i(p_i(\bar{p}_{-i}), \bar{p}_{-i})$	payoff from playing best reply on actual average price (maximum feasible payoff, cf. point B)
$\pi_i^{BRe} = \pi_i(p_i(\bar{p}^e_{-i}), \bar{p}^e_{-i})$	payoff from playing best reply on expected average price (cf. point C).
$\pi_i^{Exp} = \pi_i(p_i, \bar{p}^e_{-i})$	Expected payoff in period t (point D)
$\pi_i^{Act} = \pi_i(p_i, \bar{p}_{-i})$	Actual payoff in period t (point E)
$L_i^{Tot} = L_i^{Exp} + L_i^{BRF} + L_i^{BRF}$ $= \pi_i^* - \pi_i^{Act}$	Total loss (A -E)

Where:

$L_i^{Exp} = \pi_i^{Hyp} - \pi_i^{BRe} + \pi^{Exp} - \pi_i^{Act}$	Loss from having incorrect expectation on average price.
$L_i^{BRF} = \pi_i^{BRe} - \pi_i^{Exp}$	Loss from failure to play best reply (vertical distance between C and D)
$L_i^{ADE} = \pi_i^* - \pi_i^{Hyp}$	Loss from aggregate demand externality (vertical distance between A and B)

Numerical example. These concepts are now illustrated using the example of the SRTC post-shock payoff table. Suppose agent i of type x chooses a price of $P_{ix} = 6$, the actual average price is $\bar{P}_{-i} = 14$ and agent i expects $\bar{p}^e_{-i} = 10$. In this case, the best reply on his expectation is $P_{ix}(10) = 5$, the best reply on the actual average price is $P_{ix}(14) = 9$.

Then, his maximum feasible payoff is $\pi_{ix}^{Hyp} = 39$, his payoff from playing best reply on the expected average price is $\pi_{ix}^{BRe} = 40$, his expected payoff is $\pi_{ix}^{Exp} = 36$, and his actual payoff is $\pi_{ix}^{Act} = 18$. This agent suffers a total loss $L_{ix}^{Tot} = 40 - 18 = 22$. This loss is decomposed into the loss from failure to play best reply $L_{ix}^{BRF} = 40 - 36 = 4$, the loss from aggregate demand externality $L_{ix}^{ADE} = 40 - 39 = 1$, and the loss from having incorrect expectations $L_{ix}^{Exp} = 39 - 40 + 36 - 18 = 17$. To check, the total loss is $L_{ix}^{Tot} = 17 + 4 + 1 = 22$. Of course, the same (rounded) values are obtained when the respective values are plugged into the payoff function (1).

Results from loss decomposition. *Table* 17 shows the result of the loss decomposition in the two treatments NTC and NTS. We decompose losses for each subject i and report the averages over all subjects in the respective treatments. The entries of the table printed in italics refer to the NTC. In periods 18 to 20, i.e. before the shock, only minor differences are observed between the two treatments and full efficiency is practically reached. After the shock, efficiency losses are much larger in the complements treatment than in the substitutes treatment (this is true for periods 21-32, except for periods 25 and 26). Efficiency losses due to failure to predict average prices are much smaller in the NTS (see column 3).

Bounds to losses by virtue of parametrization. The maximum possible losses from aggregate demand externality (L^{ADE}) are relatively small by construction of the payoff matrix. They are (after the shock) 55% of equilibrium payoffs for subjects of type x and 60% for subjects of type y. This can be verified by checking the lowest payoff on the decision matrix which is on the best reply function (see appendix A2). Since the equilibrium is efficient, only losses can occur and gains from aggregate demand externality are excluded by definition. Losses from failure to play best reply on expected average price (L^{BRF}) arise when a subject fails to choose the price which maximizes payoffs given the subjective expectation for the average price in this period. Maximum losses for both types are 100% of the equilibrium payoff. Since the payoff function has a unique maximum in the decision variable P_i, all deviations from best-reply behavior yield losses, gains are excluded. Losses from having incorrect expectations (L^{Exp}) arise when the price expectation of an subject differs from actual average prices. Here, losses or gains can occur.

Table 17: Loss decomposition NTS vs. NTC (*italic*)

Period	Efficiency loss in % of equilibrium real profit (= L^{Tot}/π^*)		Efficiency loss due to failure to predict average price (= L^{Exp}/π^*)		Efficiency loss due to failure to play best reply (= L^{BRF}/π^*)		Efficiency loss due to aggregate demand externality (= L^{ADE}/π^*)	
18	0.04	*0.03*	0.01	*0.00*	0.03	*0.03*	0.00	*0.00*
19	0.02	*0.01*	0.00	*0.00*	0.01	*0.00*	0.00	*0.00*
20	0.03	*0.01*	0.01	*0.01*	0.02	*0.00*	0.00	*0.00*
21	0.18	*0.65*	0.16	*0.54*	0.02	*0.09*	0.00	*0.02*
22	0.12	*0.47*	0.09	*0.29*	0.04	*0.16*	0.00	*0.02*
23	0.11	*0.35*	0.02	*0.23*	0.07	*0.11*	0.03	*0.02*
24	0.09	*0.27*	0.07	*0.21*	0.02	*0.05*	0.00	*0.01*
25	0.19	*0.18*	0.10	*0.07*	0.06	*0.10*	0.03	*0.01*
26	0.15	*0.17*	0.08	*0.08*	0.04	*0.08*	0.03	*0.01*
27	0.09	*0.17*	0.01	*0.11*	0.07	*0.05*	0.01	*0.01*
28	0.05	*0.11*	0.02	*0.03*	0.03	*0.07*	0.00	*0.01*
29	0.04	*0.10*	0.01	*0.05*	0.02	*0.04*	0.00	*0.01*
30	0.05	*0.14*	0.02	*0.08*	0.03	*0.06*	0.00	*0.01*
31	0.01	*0.08*	0.01	*0.06*	0.01	*0.02*	0.00	*0.01*
32	0.01	*0.07*	0.00	*0.05*	0.00	*0.01*	0.00	*0.00*
33	0.08	*0.06*	0.03	*0.04*	0.03	*0.02*	0.02	*0.00*
34	0.04	*0.05*	0.01	*0.03*	0.02	*0.01*	0.00	*0.00*
35	0.02	*0.03*	-0.01	*0.02*	0.03	*0.01*	0.00	*0.00*
36	0.10	*0.10*	0.04	*0.06*	0.04	*0.04*	0.02	*0.00*
37	0.03	*0.06*	0.01	*0.05*	0.02	*0.00*	0.00	*0.00*
38	0.11	*0.14*	0.05	*0.11*	0.04	*0.03*	0.02	*0.00*
39	0.05	*0.13*	0.01	*0.10*	0.04	*0.03*	0.00	*0.00*
40	0.04	*0.03*	-0.01	*0.00*	0.05	*0.02*	0.00	*0.00*

3.3. Solving the game

Subjects were not given the functional form of the payoff function (1). Rather, the payoff function was represented in matrix form (see appendix A2). Consequently, subjects could not apply the calculus technique to find the equilibrium. Yet, they can (in principle) find the equilibrium by either A) Solving the game by iterated elimination of weakly dominated strategies, or B) by assuming that everyone chooses best replies. Note that these solution concepts do in no way depend on the nominal vs. real representation of payoffs. In the nominal treatments (NTC, NTS) we display nominal profits. That is, we multiply both sides of the profit function (1) with average nominal prices $\bar{P} \in [1, 30]$ and make sure agents understand that they are paid out according to real payoffs π at the end of the experiment.

A) Solving the game by iterated elimination of weakly dominated strategies

This procedure is explained using the example of the SRTC pre-shock tables. It is assumed that the two players of each type are symmetric.

1. step: Eliminate all weakly dominated strategies. For *x*-type players, prices in the range $16 \le P_{ix} \le 30$, for *y*-type players prices $1 \le P_{iy} \le 15$ can be eliminated.

If no player chooses dominated prices, then the average price for *x*-type players will be in the interval $11 \le \bar{P}_{-ix} \le 25$. For example, the lower bound can be found by noting that *y*-type players will at least choose $P_{iy}=$ 16 and *x*-type players at least $P_{ix}= 1$. Since $P_{-ix} = (P_x + 2P_y)/3$ we get $\bar{P}^{min}_{-ix} = (P^{min}_x + 2P^{min}_y)/3 = (1 + 2 \cdot 16) / 3 = 11$. Analogously we find that $6 \le \bar{P}_{-iy} \le 15$.

2. step: Given the intervals found in step 1, we can proceed to the second round of elimination of weakly dominated strategies. In this case we find for type *x* $16 \le P_{ix} \le 15$ and $P_{ix}= 1$, and for type *y* $16 \le P_{iy} \le 30$ and $28 \le P_{ix} \le 30$ are (weakly) dominated.

We again calculate the intervals within which average prices must lie if no player chooses the dominated strategies which have been eliminated so far. Then, we find that the respective average prices must be in the following intervals: $15 \le \bar{P}_{-ix} \le 21$, and $8 \le \bar{P}_{-iy} \le 16$.

3. step: Eliminate the prices that have become dominated by step 2. For type *x*, we find $2 \le P_{ix} \le 5$ and $P_{ix}= 10$, for *y*-type players 21 and 22 have become dominated prices.

Repeating the procedure of previous steps, we find as ranges for average prices: $17 \le \bar{P}_{-ix} \le 21$ and $12 \le \bar{P}_{-iy} \le 15$.

4. step: With the result of step 3, the prices 6 and 7 become dominated for *x*-type, and choices $13 \le P_{iy} \le 25$ are dominated for *y*-type players.

This enables us to further restrict the intervals for average prices. The remaining intervals are now $20 \le \bar{P}_{-ix} \le 21$ and $14 \le \bar{P}_{-iy} \le 15$.

5. step: With these new intervals from step 4 we find that 8 is dominated for *x*-type players, and 26 is dominated for *y*-type players.

Finally, the intervals contain only one element: $\bar{P}_{-ix} = 21, \bar{P}_{-iy} = 15$. The only prices that remain from this procedure are $P_x = 9, P_y = 27$.

These are the equilibrium prices (see *table* 16).

Thus, we found the equilibrium by iteratively eliminating (weakly) dominated strategies in 5 steps. However, the number of necessary steps varies across pre- and post-shock levels, across types and across complements vs. substitutes treatments. In any case, the equilibrium can be found in a finite number of steps.

B) Solving the game by assuming everyone plays best reply

Another possibility for subjects to find equilibrium is the following reasoning: Suppose everyone plays best reply. When can this behavior be consistent among all subjects? (This procedure is illustrated by using the same payoff tables as in the example above, i.e. SRTC pre-shock.

Suppose both x-type players choose a price of, say, $P_{ix} = 14$. Suppose both y-type players choose, say, a price of $P_{iy} = 1$. Can these choices be mutually consistent best replies? In this example, x-type players get an average price of $\bar{P}_{-ix} = 5$ [$= (14 + 2 \cdot 1) / 3 = 5.33$ rounded to the next integer], y-type players an average price of $\bar{P}_{-iy} = 10$ (9.66 rounded). The choices in the example above were not mutually consistent since the best reply for x-type on an average of $\bar{P}_{-ix} = 5$ is the choice of $P_{ix}(5) = 1$ $(\neq 14)$, a best reply for a y-type player on an average of 10 is $P_{ix}(10) = 24$. Can these choices be consistent? Now, x-type players get an average of 16, y-type players get an average of 9. Best replies on these averages are 7 for x-type, and 24 for y-type, i.e. the choices above were not mutual best replies. Given these choices, averages are 18 for x-type and 16 for y-type. Best replies on these averages are 8 for x-type, and 27 for y-type. With these choices we get averages of 21 for x-type and 14 for y-type. Best replies on these averages are 9 for x-type and 27 for y-type. The resulting (equilibrium) averages are 15 for x-type, and 21 for y-type. Finally, the choices are mutually consistent best replies, and the Nash equilibrium has been found in a finite number of steps. *Table* 18 summarizes the reasoning in this paragraph.

Wherever subjects start with this reasoning in the payoff table, they reach the unique equilibrium value. However, depending on initial values, this procedure take more or less steps of reasoning.

Table 18: Example of solving the game by assuming best reply behavior
(* denotes equilibrium values)

Period	Choice of *x*-type	Choice of *y*-type	Average of *x*-type	Average of *y*-type
1	14	1	5	10
2	1	24	16	9
3	7	24	18	16
4	8	27*	21	14
5	9*	27*	15*	21*
6	9*	27*	15*	21*

References

Agell, J., Lundborg, P. (1995): Theories of Pay and Unemployment: Survey Evidence from Swedish Manufacturing Firms. Scandinavian Journal of Economics, 97(2), 295-307

Akerlof, G. A. (1982): Labor Contracts as Partial Gift Exchange. Quarterly Journal of Economics, 97, 543-69

Akerlof, G. A.; Dickens, W. T., Perry, G. L. (1996): The Macroeconomics of Low Inflation. Brookings Papers on Economic Activity, 1, 1-59

Akerlof, G. A., Yellen, J. L. (1985a): Can Small Deviations from Rationality Make Significant Differences in Economic Equilibria? American Economic Review, 75(4), 708-20

Akerlof, G. A., Yellen, J. L. (1985b): A Near Rational Model of the Business Cycle, with Wage and Price Inertia. Quarterly Journal of Economics, 100, 823-38

Akerlof, G. A., Yellen, J. L. (1987): Rational Models of Irrational Behavior. American Economic Review, Papers and Proceedings, 77(2), 137-42

Akerlof, G. A., Yellen, J. L. (1990); The Fair Wage-Effort Hypothesis and Unemployment. Quarterly Journal of Economics, 105(May), 255-83

Alchian, A. A. (1950): Uncertainty, Evolution, and Economic Theory. Journal of Political Economy, 58, 211-21

Andersen, L. C., Jordan, J. L. (1968): Monetary and Fiscal Actions: A Test of their Relative Importance in Economic Stabilization. Federal Reserve Bank of St. Louis Review, 50(Nov.), 11-24

Andersen, T. M. (1994): Price Rigidty. Causes and Macroeconomic Consequences. Clarendon Press: Oxford

Andreoni, J. (1995): Warm-Glow versus Cold-Prickle: The Effects of Positive and Negative Framing on Cooperation in Experiments. Quarterly Journal of Economics, 60(1), 1-21

Arifovic, J. (1993): Genetic Algorithm Learning and Exchange Rate Experiments. Mimeo, McGill University

Ball, L. (1991): The Genesis of Inflation and the Costs of Disinflation. Journal of Money, Credit, and Banking, 23(Aug.), 439-52

Ball, L., Romer, D. H. (1990): Real Rigidities and the Nonneutrality of Money, Review of Economic Studies, 57(April), 183-203

Barro, R. J. (1977): Unanticipated Money Growth and Unemployment in the United States. American Economic Review, 67(1), 101-15

Barro, R. J. (1978): Unanticipated Money, Output, and the Price Level in the United States. Journal of Political Economy, 86(Aug.), 549-80

Becker, G. S. (1976): The Economic Approach to Human Behavior. University of Chicago Press: Chicago

Becker, G. S. (1993): Nobel Lecture: The Economic Way of Looking at Behavior. Journal of Political Economy, 101(3), 385-409

Belongia, M. T. (1996): Measurement Matters: Recent Results from Monetary Economics Reexamined. Journal of Political Economy, 104(5), 1065-83

Benartzi, S., Thaler, R. H. (1995): Myopic Loss Aversion and the Equity Premium Puzzle. Quarterly Journal of Economics, 110(1), 73-92

Bernanke, B. S., Carey, K. (1996): Nominal Wage Stickiness and Aggregate Supply in the Great Depression. Quarterly Journal of Economics, 111(3), 853-84

Bewley, T. F. (1995a): A Field Study on Downward Wage Rigidity. Working Paper

Bewley, T. F. (1995b): A Depressed Labor Market as Explained by Participants. American Economic Review, Papers and Proceedings, 85(2), 250-4

Blanchard, O. J. (1990): Why does Money Affect Output? A Survey. In: Friedman, B. M., Hahn, F. M. (eds.): Handbook of Monetary Economics, Vol. 2. North-Holland: Amsterdam, 779-835

Blanchard, O. J., Kiyotaki, N. (1987): Monopolistic Competition and the Effects of Aggregate Demand. American Economic Review, 77(4), 647-66

Blinder, A. S. (1987): Keynes, Lucas, and Scientific Progress. American Economic Review, Papers and Proceedings, 77(2), 130-36

Blinder, A. S. (1991): Why are Prices Sticky? Preliminary Results from an Interview Study. American Economic Review, Papers and Proceedings, 81(2), 89-100

Blinder, A. S. (1994): On Sticky Prices: Academic Theories Meet the Real World. In: Mankiw, N. G. (ed.): Monetary Policy. NBER Studies in the Business Cycle, Vol. 29. Univ. Chicago Press: Chicago, 117-54

Blinder, A. S., Choi, D. H. (1990): A Shred of Evidence on Theories of Wage Stickiness. Quarterly Journal of Economics, 105(4), 1003-15

Bomfim, A. N., Diebold, F. X. (1996): Bounded Rationality and Strategic Complementarity in a Macroeconomic Model. National Bureau of Economic Research, NBER Working paper no. 5482

Boschen, J. F., Otrok, C. M. (1994): Long-run Non-Neutrality and Superneutrality in an ARIMA Framework: Comment. American Economic Review, 84(5), 1470-3

Bulow, J. I.; Geanakoplos, J. D., Klemperer, P. D. (1985): Multimarket Oligopoly: Strategic Substitutes and Complements. Journal of Political Economy, 93(3), 488-511

Buiter, W. H., Miller, M. (1981): The Thatcher Experiment: The First Two Years. Brookings Papers on Economic Activity, 2, 315-67

Camerer, C. (1989): Bubbles and Fads in Asset Prices. Journal of Economic Surveys, 3(1), 3-41

Cameron, L. (1995): Raising the Stakes in the Ultimatum Game: Experimental Evidence form Indonesia. Discussion paper 345, Princeton University

Campbell, C. M., Kamlani, K. S. (1997): The Reasons for Wage Rigidity: Evidence from a Survey of Firms. Quarterly Journal of Economics, 112(3), 759-89

Caplin, A. S., Spulber, D. F. (1987): Menu Costs and the Neutrality of Money. Quarterly Journal of Economics, 102(Nov.), 703-25

Carlton, D. W. (1986): The Rigidity of Prices. American Economic Review, 76(4), 637-58

Cason, T., Friedman, D. (1998): A Laboratory Study of Customer Markets. Working Paper University California Santa Cruz

Cecchetti, S. G. (1986): The Frequency of Price Adjustment: A Study of Newsstand Prices of Magazines, 1953 to 1979. Journal of Econometrics, 31(April), 255-74

Churchill, N. (1982): Don't Let Inflation Get the Best of You. Harvard Business Review, March, 6-26

Coleman, W. J. (1996): Money and Output: A Test of Reverse Causation. American Economic Review, 86(1), 90-111

Conslik, J. (1996): Why Bounded Rationality? Journal of Economic Literature, 34(2), 669-700

Cooper, R. W., Haltiwanger, J. (1993): Evidence on Macroeconomic Complementarities. National Bureau of Economic Research, NBER Working Paper No. 4577

Cooper, R. W., John, A. (1988): Coordinating Coordination Failures in Keynesian Models. Quarterly Journal of Economics, 103(Aug.), 441-63

Cover, J. P. (1992): Asymmetric Effects of Positive and Negative Money Supply Shocks. Quarterly Journal of Economics, 107(Nov.), 1261-82

Daniels, B. P., Plott, C. R. (1988): Inflation Expectations in Experimental Markets. In: Tietz, R.; Albers, W., Selten, R. (eds.): Bounded Rational Behavior in Experimental Games and Markets: Proceedings of the fourth Conference on Experimental Economics, Bielefeld, Sept. 21-25 1986. Lecture Notes in Economics and Mathematical Systems, Vol. 314. Springer: New York etc., 198-218

Davis, D. D., Holt, C. A. (1993): Experimental Economics. Princeton Univ. Press: Princeton, N.J.

Diamond, P. A. (1982): Aggregate Demand Management in Search Equilibrium. Journal of Political Economy, 90(Oct.), 881-94

Diamond, P. A. (1984): A Search-Equilibrium Approach to the Micro Foundations of Macroeconomics. MIT Press: Cambridge, MA.

Driffill, J.; Mizon, G. E., Ulph, A. (1990): Costs of Inflation. In: Friedman, B. M., Hahn, F. H. (eds.): Handbook of Monetary Economics, Vol. 2. North-Holland: Amsterdam, Ch. 19, 1013-66

Duffy, J. (1998): Monetary Theory in the Laboratory. Federal Reserve Bank of St. Louis Review, September/October, 9-26

Dusansky, R. (1980): Utility Function Transformations and Money Illusion: Reply and Further Results. American Economic Review, 70(4), 823-5

Eatwell, J.; Millgate, M., Newman, P. (eds.) (1989): Money (A collection of articles from The New Palgrave). W. W. Norton: New York, London

Einhorn, H. J., Hogarth, R. M. (1978): Confidence in Judgment: Persistence in the Illusion of Validity. Psychological Review, 85(5), 395-416

Elster, J. (1984): Ulysses and the Sirens. Cambridge Univ. Press.: Cambridge

Enders, W., Falk, B. (1984): A Microeconomic Test of Monetary Neutrality. Review of Economics and Statistics, 66(4), 666-9

Evans, G. W., Ramey, G. (1992): Expectation Calculation and Macroeconomic Dynamics. American Economic Review, 82(1), 207-24

Fair, R. (1971): Labor Force Participation, Wage Rates and Money Illusion. Review of Economics and Statistics, 53(2), 164-8

Falk, A., Tyran, J.-R. (1997): Experimentelle Wirtschaftsforschung. Wirtschaftswissenschaftliches Studium WiSt, 6, 326-329

Fehr, E., Falk, A. (1998): Wage Rigidities in a Competitive Incomplete Contracts Market - An Experimental Investigation. Forthcoming in: Journal of Political Economy, 107(1), 106-34

Fehr, E.; Kirchsteiger, G., Riedl, A. (1993): Does Fairness prevent Market Clearing? An Experimental Investigation. Quarterly Journal of Economics, 108(2), 437-60

Fehr, E.; Kirchsteiger, G., Riedl, A. (1998): Gift Exchange and Reciprocity in Competitive Experimental Markets. European Economic Review 42, 1-34

Fehr, E., Schmidt, K. (1998): A Theory of Fairness, Competition and Cooperation. Working Paper, Institute for Empirical Research in Economics, University of Zurich

Fehr, E., Tougareva, E. (1995): Do Competitive Markets with High Stakes Remove Reciprocal Fairness? Working Paper, Institute for Empirical Research in Economics, University of Zurich

Feige, E. L., Pearce, D. K. (1979): The Causal Relationship Between Money and Income: Some Caveats for Time Series Analysis. Review of Economics and Statistics, 61, 521-33

Feldstein, M. (1983): Inflation, Tax Rules, and Capital Formation. University of Chicago Press: Chicago

Feldstein, M. (1997): The Costs and Benefits of Going from Low Inflation to Price Stability. In: Romer, C. D., Romer, D. H. (eds.), Reducing Inflation. National Bureau of Economic Research Studies in the Business Cycle, Vol. 30. University of Chicago Press: Chicago

Fiedler, U. (1979): Experimentelle Untersuchung eines geschlossenen gesamtwirtschaftlichen Systems. Dissertation Berlin, Darmstadt

Fischer, S. (1977): Long-Term Contracts, Rational Expectations, and the Optimal Money Supply Rule. Journal of Political Economy, 85(Feb.), 191-205

Fischer S., Modigliani, F. (1986): Towards an Understanding of the Real Effects and Costs of Inflation. Weltwirtschaftliches Archiv, 114, 810-33

Fisher, I. (1922): The Purchasing Power of Money, 2nd ed. Macmillan: New York

Fisher, I. (1928): The Money Illusion. Longmans: Toronto

Fisher, M., Seater, J. (1993): Long-run Neutrality and Superneutrality in an ARIMA Framework. American Economic Review, 83(3), 402-15

Franciosi, R.; Kujal, P.; Michelitsch, R.; Smith, V. L., Deng, G. (1995): Fairness: Effect on Temporary and Equilibrium Prices in Posted-offer Markets. Economic Journal, 105(431), 938-50

Frey, B. S. (ed.) (1997): Political Business Cycles. Elgar Publishers: Cheltenham

Frey, B. S., Eichenberger, R. (1994): Economic Incentives Transform Psychological Anomalies. Journal of Economic Behavior and Organization, 23(2), 215-34

Friedman, B. M., Hahn, F. M. (eds.) (1990): Handbook of Monetary Economics, Vol. 2. North-Holland: Amsterdam

Friedman, B. M., Kuttner, K. N. (1992): Money, Income, Prices, and Interest Rates. American Economic Review, 85(3), 472-92

Friedman, B. M., Kuttner, K. N. (1993): Another Look at the Money-Income Causality. Journal of Econometrics, 57(May/June), 189-203

Friedman, D. (1998): Monty Hall's Three Doors: Construction and Deconstruction of a Choice Anomaly. American Economic Review, 88(4), 933-46

Friedman, D., Sunder, S. (1994): Experimental Methods. Cambridge University Press: Cambridge

Friedman, M. (1953): Essays in Positive Economics. University of Chicago Press: Chicago

Friedman, M. (1989): The Quantity Theory of Money. In: Eatwell, J.; Milgate, M., Newman, P. (eds.): Money. W.W. Norton: New York, London, 1-40

Friedman, M., Schwartz, A. J. (1963): A Monetary History of the United States, 1867 - 1960. Princeton Univ. Press for the National Bureau of Economic Research: Princeton

Gode, D. K., Sunder, S. (1993): Allocative Efficiency of Markets with Zero-Intelligence Traders: Markets as Partial Substitute for Individual Rationality. Journal of Political Economy, 101(1), 119-37

Gordon, R. J. (1990): What Is New Keynesian Economics? Journal of Economic Literature, 28(Sept.), 1115-71

Granger, C. W. J. (1969): Investigating Causal Relations by Econometric Models and Cross-Spectral Methods. Econometrica, 37(July), 424-38

Grether, D. (1980): Bayes Rule as a Descriptive Model: The Representativeness Heuristic. Quarterly Journal of Economics, 95(Nov.), 537-57

Grether, D., Plott, C. (1979): Economic Theory of Choice and the Preference Reversal Phenomenon. American Economic Review, 69(4), 622-38

Groshen, E. L., Schweitzer, M. E. (1997): Identifying Inflation's Grease and Sand Effects in the Labor Market. Federal Reserve Bank of New York, 31(Oct.)

Gustafson, E., Hadley, L. (1989): A Dynamic Simultaneous Equations Model. Quarterly Review of Economics and Business, 29(2), 63-75

Güth, W.; Schmittberger, R., Schwarze, B. (1982): An Experimental Analysis of Ultimatum Bargaining. Journal of Economic Behavior and Organization, 3, 367-88

Haltiwanger, J., Waldman, M. (1985): Rational Expectations and the Limits of Rationality: An Analysis of Heterogeneity. American Economic Review, 75(3), 326-40

Haltiwanger, J., Waldman, M. (1989): Rational Expectations and Strategic Complements: The Implications for Macroeconomics. Quarterly Journal of Economics, 104(August), 463-84

Hart, O. (1982): A Model of Imperfect Competition with Keynesian Features. Quarterly Journal of Economics, 97(1), 109-38

Hey, J. D., DiCagno, D. (1998): Sequential Markets: An Experimental Investigation of Clower's Dual-Decision Hypothesis. Experimental Economics, 1, 63-85

Ho, T.-H.; Camerer, C., Weigelt, K. (1998): Iterated Dominance and Interated Best Response in Experimental „p-Beauty Contests". American Economic Review, 88(4), 947-69

Howitt, P. (1985): Transaction Costs and the Theory of Unemployment. American Economic Review, 75(1), 88-100

Howitt, P. (1989): Money Illusion. In: Eatwell, J.; Milgate, M., Newman, P. (eds.): Money. W.W. Norton: New York, London, 244-7

Howitt, P., Patinkin, D. (1980): Utility Function Transformations and Money Illusion: Comments. American Economic Review, 70(4), 819-22, 826-8

Hume, D. (1752): Of Money; Of Interest. Both essays are reprinted in: Rotwein, E. (ed.) (1970): Writings on Economics. Univ. of Wisconsin Press: Madison

Jones, S. R. G., Stock, J. H. (1987): Demand Disturbances and Aggregate Fluctuations: The Implications of Near Rationality. Economic Journal, 97(March), 49-64

Kachelmeier, S. J., Shehata, M. (1992): Examining Risk Preferences under High Monetary Incentives: Experimental Evidence from the People's Republic of China. American Economic Review, 82(5), 1120-41

Kagel, J. H., Roth, A. E. (eds.) (1995): The Handbook of Experimental Economics. Princeton Univ. Press: Princeton, N.J.

Kahn, S. (1997): Evidence of Nominal Wage Stickiness from Microdata. American Economic Review, 87(5), 993-1008

Kahneman, D., Tversky, A. (1979): Prospect Theory: An Analysis of Decision Under Risk. Econometrica, 47, 263-91

Kahneman, D.; Knetsch, J., Thaler, R. (1986): Fairness as a Constraint on Profit Seeking: Entitlements in the Market. American Economic Review, 76(4), 728-41

Kashyap, A. K. (1987): Sticky Prices: New Evidence from Retail Catalogues. Finance and Economics Discussion Series, Federal Reserve Board: Washington, D.C.

Keynes, J. M. (1923): A Tract on Monetary Reform. Macmillan: London

Keynes, J. M. (1936): The General Theory of Employment, Interest, and Money. Macmillan: London

King, R. G. (1981): Monetary Information and Monetary Neutrality. Journal of Monetary Economics, 7(March), 195-206

King, R. G., Plosser, C. I. (1984): Money, Credit, and Prices in a Real Business Cycle. American Economic Review, 74(3), 363-80

Kirchkamp, O., Bernasconi, M. (1998): Why Monetary Policy Matters - An Experimental Study of Saving, Inflation and Monetary Policies in an Overlapping Generations Model. Working Paper University of Mannheim, SFB 504

Kiyotaki, N. (1988): Implications of Multiple Expectational Equilibria under Monopolistic Competition. Mimeo, May 1985

Kormendi, R. C., Meguire, P. G. (1984): Cross Regime Evidence of Macroeconomic Rationality. Journal of Political Economy, 92(5), 875-908

Kreps, D. M.; Milgrom, P.; Roberts, J., Wilson, R. (1982): Rational Cooperation in the Finitely Repeated Prisoners' Dilemma. Journal of Economic Theory, 27(2), 245-52

Kretzmer, P. E. (1989): The Cross-Industry Effects of Unanticipated Money in an Equilibrium Business Cycle Model. Journal of Monetary Economics, 23(2), 275-96

Krugman, P. (1994): Peddling Prosperity. W.W. Norton: New York, London

Kydland, F. E., Prescott, E. C. (1990): Business Cycles: Real Facts and Monetary Myth. Federal Bank Reserve of Minneapolis Quarterly Review, 14(Spring), 3-18

Kydland, F. E., Prescott, E. C. (1996): The Computational Experiment: An Econometric Tool. Journal of Economic Perspectives, 10(1), 69-85

Langdana, F. K. (1994): An Experimental Verification of the Lucas "Islands" Approach to Business Cycles. Journal of Economic Behavior and Organization, 25(2), 271-80

Ledyard, J. O. (1995): Public Goods: A Survey of Experimental Research. In: Kagel, J. H., Roth, A. E. (eds.): Handbook of Experimental Economics. Princeton University Press, Princeton

Leijonhufvud, A. (1981): Information and Coordination: Essays in Macroeconomic Theory. Oxford Univ. Press, New York

Leontief, W. (1936): The Fundamental Assumptions of Mr. Keynes' Monetary Theory of Unemployment. Quarterly Journal of Economics, 5(Nov.), 192-7

Lian, P., Plott, C. R. (1994): General Equilibrium, Markets, Macroeconomics and Money in a Laboratory Experimental Environment. Caltech Social Science Working Paper 842

Loewenstein, G., Sicherman, N. (1991): Do Workers Prefer Increasing Wage Profiles? Journal of Labor Economics, 9(1), 67-84

Lovell, M. C. (1986): Tests of the Rational Expectations Hypothesis. American Economic Review, 76(1), 110-24

Lucas, R. E. (1972): Expectations and the Neutrality of Money. Journal of Economic Theory, 4(April), 103-24

Lucas, R. E. (1973): Some International Evidence on Output-Inflation Trade-offs. American Economic Review, 63(3), 326-34

Lucas, R. E. (1975): An Equilibrium Model of the Real Business Cycle. Journal of Political Economy, 83(Dec.), 1113-44

Lucas, R. E. (1976): Econometric Policy Evaluation: A Critique. Carnegie-Rochester Conference Series on Public Policy, 1, 19-46

Lucas, R. E. (1996): Nobel Lecture: Monetary Neutrality. Journal of Political Economy, 104(4), 661-82

Mailath, G. J. (1998): Do People Play Nash Equilibrium? Lessons from Evolutionary Game Theory. Journal of Economic Literature, 36(3), 1347-74

Mankiw, N. G. (1985): Small Menu Costs and Large Business Cycles: A Macroeconomic Model of Monopoly. Quarterly Journal of Economics, 100(May), 529-37

Mankiw, N. G. (1990): A Quick Refresher Course in Macroeconomics. Journal of Economic Literature, 28(Dec.), 1645-60

Mankiw, N. G., Romer, D. H. (eds.) (1991): New Keynesian Economics, Vol. 1&2. MIT Press: Cambridge, MA.

Marimon, R.; Spear, S., Sunder, S. (1991): Expectationally-driven Market Volatility: An Experimental Study. Journal of Economic Theory, 61, 74-103

Marimon, R., Sunder, S. (1993): Indeterminacy of Equilibria in a Hyperinflationary World: Experimental Evidence. Econometrica, 61(5), 1073-107

Mas, I. (1995): Things Governments Do to Money: A Recent History of Currency Reform Schemes and Scams. Kyklos, 48(4), 483-512

McCabe, K. A. (1989): Fiat Money as a Store of Value in an Experimental Market. Journal of Economic Behavior and Organization, 12, 215-31

McCandless, G. T.; Weber, W. E. (1995): Some Monetary Facts. Federal Reserve Bank of Minneapolis Quarterly Review, 19(3), 2-11

McNeil, B. J.; Pauker, S. G., Tversky, A. (1988): On the Framing of Medical Decisions. In: Bell, D.; Raiffa, H., Tversky, A. (eds.): Decision Making: Descriptive, Normative, and Prescriptive Interactions. Cambridge University Press: New York

Niemi, B., Lloyd, C. (1981): Female Labor Supply in the Context of Inflation. American Economic Review, Papers and Proceedings, 71(2), 70-2

Noussair, C. W.; Plott, C. R., Rietzman, R. G. (1995): The Principles of Exchange Rate Determination in an International Finance Experiment. Journal of Political Economy, 105(4), 822-61

Milgrom, P., Roberts, J. (1990): Rationalizability, Learning, and Equilibrium in Games with Strategic Complementarities. Econometrica, 58(6), 1255-77

Mishkin, F. (1983): A Rational Expectations Approach to Macroeconometrics. NBER and the University of Chicago Press: Chicago

Mishkin, F. S. (1995): Symposium on the Monetary Transmission Mechanism. Journal of Economic Perspectives, 9(4), 3-10

Modigliani, F., Cohn, R. (1979): Inflation, Rational Valuation, and the Market. Financial Analysts Journal, 35(March), 24-44

Montesano, A. (1981): The Notion of Money Illusion. Rivista di Politica Economica, 71(0), 139-47

Moulton, B. R. (1996): Bias in the Consumer Price Index: What is the Evidence? Journal of Economic Perspectives, 10(4), 159-77

Nagel, R. (1995): Unraveling in Guessing Games: An Experimental Study. American Economic Review, 85(5), 1313-26

Naish, H. F. (1988): Imperfect Competition, Price Adjustment Costs, and the Long-Run Phillips Curve. Journal of Macroeconomics, 10(1), 103-24

Naish, H. F. (1993): The Near Optimality of Adaptive Expectations. Journal of Economic Behavior and Organization, 20, 3-22

Ochs, J. (1995): Coordination Problems. In: Kagel, J. H., Roth, A. E. (eds.): The Handbook of Experimental Economics. Princeton Univ. Press: Princeton, N.J., Chap. 3, 195-251

Oh, S., Waldman, M. (1994): Strategic Complementarity Slows Macroeconomic Adjustment to Temporary Shocks. Economic Inquiry, 32(April), 318-29

Okun, A. M. (1975): Inflation: It's Mechanics and Welfare Costs. Brookings Papers on Economic Activity, 2, 351-90

Olekalns, N. (1996): Some further Evidence on the Long-run Neutrality of Money. Economics Letters, 50(2), 393-398

Patinkin, D. (1949): The Indeterminancy of Absolute Prices in Classical Economic Theory. Econometrica, 17(1), 1-27

Patinkin, D. (1965): Money, Interest, and Prices, 2nd ed. New York: Harper and Row

Patinkin, D. (1989): Neutrality of Money. In: Eatwell, J.; Milgate, M., Newman, P. (eds.): Money. W.W. Norton: New York, London, 1-40

Patinkin, D., Steiger, O. (1989): In Search of the "Veil of Money" and the "Neutrality of Money": A Note on the Origin of Terms. Scandinavian Journal of Economics, 91(1), 131-46

Phelps, E. S. (1970): Microeconomic Foundations of Employment and Inflation Theory. Macmillan: London

Pindyk, R. S., Rubinfeld, D. L. (1991): Econometric Models and Economic Forecasts. McGraw-Hill: New York

Rabin, M. (1996): Psychology and Economics. Working Paper Univ. of California - Berkeley, Sept. 1996

Romer, C. D., Romer, D. H. (1989): Does Monetary Policy Matter? A New Test in the Spirit of Friedman and Schwartz. NBER Macroeconomics Annual, 4, 121-70

Romer, D. H. (1996): Advanced Macroeconomics. McGraw-Hill: New York, London

Rotemberg, J. J. (1993): Monetary Aggregates, Monetary Policy and Economic Activity: Commentary. Federal Reserve Bank of St. Louis Review, 75(March), 36-41

Russell, T., Thaler, R. H. (1985): The Relevance of Quasi-Rationality in Competitive Markets. American Economic Review, 75(5), 1071-82. Reprinted in: Thaler, R. H. (1994): Quasi Rational Economics. Russel Sage, New York, 239-57

Samuelson, P. A. (1983): Foundations of Economic Analysis. Harvard University Press, Cambridge, MA.

Samuelson, P. A., Nordhaus, W. D. (1985): Economics, 12th ed. McGraw-Hill: New York etc.

Samuelson, P. A., Nordhaus, W. D. (1992): Economics, 14th ed. McGraw-Hill: New York etc.

Sargent, T. J. (1976): A Classical Macroeconometric Model for the United States. Journal of Political Economy, 84(April), 207-37

Sargent, T. J. (1982): The End of Four Big Inflations. In: Hall, R.E. (ed.): Inflation. Chicago University Press: Chicago, 41-98

Sargent, T. J. (1993): Bounded Rationality in Macroeconomics. Oxford University Press

Sehti, R., Franke, R. (1995): Behavioural Heterogeneity under Evolutionary Pressure: Macroeconomic Implications of Costly Optimisation. Economic Journal, 105(430), 583-600

Selten, R., Berg, C. C. (1970): Drei experimentelle Oligopolserien mit kontinuierlichem Zeitablauf. In: Sauermann, H. (ed.): Beiträge zur experimentellen Wirtschaftsforschung, Vol. II. Tübingen, 162-221

Shafir, E.; Diamond, P., Tversky, A. (1997): On Money Illusion. Quarterly Journal of Economics, 112(2), 341-74

Shafir, E., Thaler, R. H. (1995): Mental Accounting Through Time. Manuscript

Shapiro, M. D., Wilcox, D. W. (1996): Mismeasurement in the Consumer Price Index: An Evaluation. National Bureau of Economic Research, NBER Working Paper No. 5590

Shiller, R. J. (1984): Stock Prices and Social Dynamics. Brookings Papers on Economic Activity, 2, 457-510

Shiller, R. J. (1997): Why do People Dislike Inflation? In: Romer, C. D., Romer, D. H. (eds.): Reducing Inflation. Motivation and Strategy. National Bureau of Economic Research Studies in Business Cycles, 30, 13-70

Sims, C. A. (1972): Money, Income, and Causality. American Economic Review, 62(4), 540-52

Smith, V. L. (1976): Experimental Economics: Induced Value Theory. American Economic Review, Papers and Proceedings, 66(2), 274-9

Smith, V. L. (1982): Microeconomic Systems as an Experimental Science. American Economic Review, 72(5), 923-55

Smith, V. L. (1989): Theory, Experiment and Economics. Journal of Economic Perspectives, 90(4), 151-69

Smith, V. L. (1991): Rational choice: The Contrast between Economics and Psychology. Journal of Political Economy, 90(4), 877-97

Smith, V. L.; Suchanek, G. L., Williams, A. W. (1988): Bubbles, Crashes, and Endogenous Expectations in Experimental Spot Asset Markets. Econometrica, 56(6), 1119-52

Stigler, G. (1966): The Theory of Price. Macmillan: New York

Stock, J. H., Watson, M. W. (1989): Interpreting the Evidence on Money Income-Causality. Journal of Econometrics, 40(Jan.), 161-81

Summers, L. (1988): Relative Wages, Efficiency Wages, and Keynesian Unemployment. American Economic Review, Papers and Proceedings, 78(2), 382-8

Sunder, S. (1995): Experimental Asset Markets. In: Kagel, J. H., Roth, A. E. (eds.): The Handbook of Experimental Economics. Princeton Univ. Press: Princeton, N.J., Chap. 6, 445-501

Taylor, J. B. (1979): Staggered Wage Setting in a Macro Model. American Economic Review, 69(2), 108-13

Taylor, J. B. (1997): A Core of Practical Macroeconomics. American Economic Review, Papers and Proceedings, 87(2), 233-5

Thaler, R. H., Tversky, A. (1996): Myopic Loss Aversion in Financial Investment: An Experimental Study. Unpublished Manuscript, University of Chicago

Thurman, W. N., Fisher, M. E. (1988): Chickens, Eggs, and Causality, or Which Came First? American Journal of Agricultural Economics, May, 237-8

Tietz, R. (1972): The Macroeconomic Experimental Game KRESKO - Experimental Design and the Influence of Economic Knowledge on Decision Behavior. In: Sauermann, H. (ed.): Beiträge zur experimentellen Wirtschaftsforschung, Vol. 3. Tübingen, 267-88

Tietz, R. (1996): Experimentelle Wirtschaftsforschung - Wege zur Modellierung eingeschränkter Rationalität. Ökonomie und Gesellschaft, 13, Campus Verlag: New York, 120-55

Tobin, J. (1970): Post Hoc ergo Propter Hoc? Quarterly Journal of Economics, 84, 310-7

Tobin, J. (1972): Inflation and Unemployment. American Economic Review, 62, 1-18

Tobin, J. (1989): On the Theory of Macroeconomic Policy. Cowles Foundation discussion paper No. 931

Tversky, A., Kahneman, D. (1986): Rational Choice and the Framing of Decisions. Journal of Business, 49(4), 251-78

Van Huyck, J. B.; Battalio, R. C., Beil, R. O. (1990): Tacit Coordination Games, Strategic Uncertainty and Coordination Failure. American Economic Review, 80(1), 234-48

Weil, P. (1989): Increasing Returns and Animal Spirits. American Economic Review, 79(4), 889-94

Wilson, B. J. (1998): Menu Costs and Nominal Price Friction: An Experimental Examination. Journal Economic Behavior and Organization, 35(3), 371-88.

Woltjer, G. (1996): Experimental Macroeconomics. Ökonomie und Gesellschaft, 13, Campus Verlag: New York, 156-85

List of tables

List of figures

Author index

Vol. 377: A. Villar, Operator Theorems with Applications to Distributive Problems and Equilibrium Models. XVI, 160 pages. 1992.

Vol. 378: W. Krabs, J. Zowe (Eds.), Modern Methods of Optimization. Proceedings, 1990. VIII, 348 pages. 1992.

Vol. 379: K. Marti (Ed.), Stochastic Optimization. Proceedings, 1990. VII, 182 pages. 1992.

Vol. 380: J. Odelstad, Invariance and Structural Dependence. XII, 245 pages. 1992.

Vol. 381: C. Giannini, Topics in Structural VAR Econometrics. XI, 131 pages. 1992.

Vol. 382: W. Oettli, D. Pallaschke (Eds.), Advances in Optimization. Proceedings, 1991. X, 527 pages. 1992.

Vol. 383: J. Vartiainen, Capital Accumulation in a Corporatist Economy. VII, 177 pages. 1992.

Vol. 384: A. Martina, Lectures on the Economic Theory of Taxation. XII, 313 pages. 1992.

Vol. 385: J. Gardeazabal, M. Regúlez, The Monetary Model of Exchange Rates and Cointegration. X, 194 pages. 1992.

Vol. 386: M. Desrochers, J.-M. Rousseau (Eds.), Computer-Aided Transit Scheduling. Proceedings, 1990. XIII, 432 pages. 1992.

Vol. 387: W. Gaertner, M. Klemisch-Ahlert, Social Choice and Bargaining Perspectives on Distributive Justice. VIII, 131 pages. 1992.

Vol. 388: D. Bartmann, M. J. Beckmann, Inventory Control. XV, 252 pages. 1992.

Vol. 389: B. Dutta, D. Mookherjee, T. Parthasarathy, T. Raghavan, D. Ray, S. Tijs (Eds.), Game Theory and Economic Applications. Proceedings, 1990. IX, 454 pages. 1992.

Vol. 390: G. Sorger, Minimum Impatience Theorem for Recursive Economic Models. X, 162 pages. 1992.

Vol. 391: C. Keser, Experimental Duopoly Markets with Demand Inertia. X, 150 pages. 1992.

Vol. 392: K. Frauendorfer, Stochastic Two-Stage Programming. VIII, 228 pages. 1992.

Vol. 393: B. Lucke, Price Stabilization on World Agricultural Markets. XI, 274 pages. 1992.

Vol. 394: Y.-J. Lai, C.-L. Hwang, Fuzzy Mathematical Programming. XIII, 301 pages. 1992.

Vol. 395: G. Haag, U. Mueller, K. G. Troitzsch (Eds.), Economic Evolution and Demographic Change. XVI, 409 pages. 1992.

Vol. 396: R. V. V. Vidal (Ed.), Applied Simulated Annealing. VIII, 358 pages. 1992.

Vol. 397: J. Wessels, A. P. Wierzbicki (Eds.), User-Oriented Methodology and Techniques of Decision Analysis and Support. Proceedings, 1991. XII, 295 pages. 1993.

Vol. 398: J.-P. Urbain, Exogeneity in Error Correction Models. XI, 189 pages. 1993.

Vol. 399: F. Gori, L. Geronazzo, M. Galeotti (Eds.), Nonlinear Dynamics in Economics and Social Sciences. Proceedings, 1991. VIII, 367 pages. 1993.

Vol. 400: H. Tanizaki, Nonlinear Filters. XII, 203 pages. 1993.

Vol. 401: K. Mosler, M. Scarsini, Stochastic Orders and Applications. V, 379 pages. 1993.

Vol. 402: A. van den Elzen, Adjustment Processes for Exchange Economies and Noncooperative Games. VII, 146 pages. 1993.

Vol. 403: G. Brennscheidt, Predictive Behavior. VI, 227 pages. 1993.

Vol. 404: Y.-J. Lai, Ch.-L. Hwang, Fuzzy Multiple Objective Decision Making. XIV, 475 pages. 1994.

Vol. 405: S. Komlósi, T. Rapcsák, S. Schaible (Eds.), Generalized Convexity. Proceedings, 1992. VIII, 404 pages. 1994.

Vol. 406: N. M. Hung, N. V. Quyen, Dynamic Timing Decisions Under Uncertainty. X, 194 pages. 1994.

Vol. 407: M. Ooms, Empirical Vector Autoregressive Modeling. XIII, 380 pages. 1994.

Vol. 408: K. Haase, Lotsizing and Scheduling for Production Planning. VIII, 118 pages. 1994.

Vol. 409: A. Sprecher, Resource-Constrained Project Scheduling. XII, 142 pages. 1994.

Vol. 410: R. Winkelmann, Count Data Models. XI, 213 pages. 1994.

Vol. 411: S. Dauzère-Péres, J.-B. Lasserre, An Integrated Approach in Production Planning and Scheduling. XVI, 137 pages. 1994.

Vol. 412: B. Kuon, Two-Person Bargaining Experiments with Incomplete Information. IX, 293 pages. 1994.

Vol. 413: R. Fiorito (Ed.), Inventory, Business Cycles and Monetary Transmission. VI, 287 pages. 1994.

Vol. 414: Y. Crama, A. Oerlemans, F. Spieksma, Production Planning in Automated Manufacturing. X, 210 pages. 1994.

Vol. 415: P. C. Nicola, Imperfect General Equilibrium. XI, 167 pages. 1994.

Vol. 416: H. S. J. Cesar, Control and Game Models of the Greenhouse Effect. XI, 225 pages. 1994.

Vol. 417: B. Ran, D. E. Boyce, Dynamic Urban Transportation Network Models. XV, 391 pages. 1994.

Vol. 418: P. Bogetoft, Non-Cooperative Planning Theory. XI, 309 pages. 1994.

Vol. 419: T. Maruyama, W. Takahashi (Eds.), Nonlinear and Convex Analysis in Economic Theory. VIII, 306 pages. 1995.

Vol. 420: M. Peeters, Time-To-Build. Interrelated Investment and Labour Demand Modelling. With Applications to Six OECD Countries. IX, 204 pages. 1995.

Vol. 421: C. Dang, Triangulations and Simplicial Methods. IX, 196 pages. 1995.

Vol. 422: D. S. Bridges, G. B. Mehta, Representations of Preference Orderings. X, 165 pages. 1995.

Vol. 423: K. Marti, P. Kall (Eds.), Stochastic Programming. Numerical Techniques and Engineering Applications. VIII, 351 pages. 1995.

Vol. 424: G. A. Heuer, U. Leopold-Wildburger, Silverman's Game. X, 283 pages. 1995.

Vol. 425: J. Kohlas, P.-A. Monney, A Mathematical Theory of Hints. XIII, 419 pages, 1995.

Vol. 426: B. Finkenstädt, Nonlinear Dynamics in Economics. IX, 156 pages. 1995.

Vol. 427: F. W. van Tongeren, Microsimulation Modelling of the Corporate Firm. XVII, 275 pages. 1995.

Vol. 428: A. A. Powell, Ch. W. Murphy, Inside a Modern Macroeconometric Model. XVIII, 424 pages. 1995.

Vol. 429: R. Durier, C. Michelot, Recent Developments in Optimization. VIII, 356 pages. 1995.

Vol. 430: J. R. Daduna, I. Branco, J. M. Pinto Paixão (Eds.), Computer-Aided Transit Scheduling. XIV, 374 pages. 1995.

Vol. 431: A. Aulin, Causal and Stochastic Elements in Business Cycles. XI, 116 pages. 1996.

Vol. 432: M. Tamiz (Ed.), Multi-Objective Programming and Goal Programming. VI, 359 pages. 1996.

Vol. 433: J. Menon, Exchange Rates and Prices. XIV, 313 pages. 1996.

Vol. 434: M. W. J. Blok, Dynamic Models of the Firm. VII, 193 pages. 1996.

Vol. 435: L. Chen, Interest Rate Dynamics, Derivatives Pricing, and Risk Management. XII, 149 pages. 1996.

Vol. 436: M. Klemisch-Ahlert, Bargaining in Economic and Ethical Environments. IX, 155 pages. 1996.

Vol. 437: C. Jordan, Batching and Scheduling. IX, 178 pages. 1996.

Vol. 438: A. Villar, General Equilibrium with Increasing Returns. XIII, 164 pages. 1996.

Vol. 439: M. Zenner, Learning to Become Rational. VII, 201 pages. 1996.

Vol. 440: W. Ryll, Litigation and Settlement in a Game with Incomplete Information. VIII, 174 pages. 1996.

Vol. 441: H. Dawid, Adaptive Learning by Genetic Algorithms. IX, 166 pages.1996.

Vol. 442: L. Corchón, Theories of Imperfectly Competitive Markets. XIII, 163 pages. 1996.

Vol. 443: G. Lang, On Overlapping Generations Models with Productive Capital. X, 98 pages. 1996.

Vol. 444: S. Jørgensen, G. Zaccour (Eds.), Dynamic Competitive Analysis in Marketing. X, 285 pages. 1996.

Vol. 445: A. H. Christer, S. Osaki, L. C. Thomas (Eds.), Stochastic Modelling in Innovative Manufactoring. X, 361 pages. 1997.

Vol. 446: G. Dhaene, Encompassing. X, 160 pages. 1997.

Vol. 447: A. Artale, Rings in Auctions. X, 172 pages. 1997.

Vol. 448: G. Fandel, T. Gal (Eds.), Multiple Criteria Decision Making. XII, 678 pages. 1997.

Vol. 449: F. Fang, M. Sanglier (Eds.), Complexity and Self-Organization in Social and Economic Systems. IX, 317 pages, 1997.

Vol. 450: P. M. Pardalos, D. W. Hearn, W. W. Hager, (Eds.), Network Optimization. VIII, 485 pages, 1997.

Vol. 451: M. Salge, Rational Bubbles. Theoretical Basis, Economic Relevance, and Empirical Evidence with a Special Emphasis on the German Stock Market.IX, 265 pages. 1997.

Vol. 452: P. Gritzmann, R. Horst, E. Sachs, R. Tichatschke (Eds.), Recent Advances in Optimization. VIII, 379 pages. 1997.

Vol. 453: A. S. Tangian, J. Gruber (Eds.), Constructing Scalar-Valued Objective Functions. VIII, 298 pages. 1997.

Vol. 454: H.-M. Krolzig, Markov-Switching Vector Autoregressions. XIV, 358 pages. 1997.

Vol. 455: R. Caballero, F. Ruiz, R. E. Steuer (Eds.), Advances in Multiple Objective and Goal Programming. VIII, 391 pages. 1997.

Vol. 456: R. Conte, R. Hegselmann, P. Terna (Eds.), Simulating Social Phenomena. VIII, 536 pages. 1997.

Vol. 457: C. Hsu, Volume and the Nonlinear Dynamics of Stock Returns. VIII, 133 pages. 1998.

Vol. 458: K. Marti, P. Kall (Eds.), Stochastic Programming Methods and Technical Applications. X, 437 pages. 1998.

Vol. 459: H. K. Ryu, D. J. Slottje, Measuring Trends in U.S. Income Inequality. XI, 195 pages. 1998.

Vol. 460: B. Fleischmann, J. A. E. E. van Nunen, M. G. Speranza, P. Stähly, Advances in Distribution Logistic. XI, 535 pages. 1998.

Vol. 461: U. Schmidt, Axiomatic Utility Theory under Risk. XV, 201 pages. 1998.

Vol. 462: L. von Auer, Dynamic Preferences, Choice Mechanisms, and Welfare. XII, 226 pages. 1998.

Vol. 463: G. Abraham-Frois (Ed.), Non-Linear Dynamics and Endogenous Cycles. VI, 204 pages. 1998.

Vol. 464: A. Aulin, The Impact of Science on Economic Growth and its Cycles. IX, 204 pages. 1998.

Vol. 465: T. J. Stewart, R. C. van den Honert (Eds.), Trends in Multicriteria Decision Making. X, 448 pages. 1998.

Vol. 466: A. Sadrieh, The Alternating Double Auction Market. VII, 350 pages. 1998.

Vol. 467: H. Hennig-Schmidt, Bargaining in a Video Experiment. Determinants of Boundedly Rational Behavior. XII, 221 pages. 1999.

Vol. 468: A. Ziegler, A Game Theory Analysis of Options. XIV, 145 pages. 1999.

Vol. 469: M. P. Vogel, Environmental Kuznets Curves. XIII, 197 pages. 1999.

Vol. 470: M. Ammann, Pricing Derivative Credit Risk. XII, 228 pages. 1999.

Vol. 471: N. H. M. Wilson (Ed.), Computer-Aided Transit Scheduling. XI, 444 pages. 1999.

Vol. 472: J.-R. Tyran, Money Illusion and Strategic Complementarity as Causes of Monetary Non-Neutrality. X, 228 pages. 1999.

Vol. 473: S. Helber, Performance Analysis of Flow Lines with Non-Linear Flow of Material. IX, 280 Seiten. 1999.

GPSR Compliance

The European Union's (EU) General Product Safety Regulation (GPSR) is a set of rules that requires consumer products to be safe and our obligations to ensure this.

If you have any concerns about our products, you can contact us on ProductSafety@springernature.com

In case Publisher is established outside the EU, the EU authorized representative is:

Springer Nature Customer Service Center GmbH
Europaplatz 3
69115 Heidelberg, Germany

Batch number: 10844675

Printed by Printforce, the Netherlands